THE OAK PAPERS

Also by James Canton

Ancient Wonderings: Journeys Into Prehistoric Britain
Out of Essex: Re-Imagining a Literary Landscape
From Cairo to Baghdad: British Travellers in Arabia

The Oak Papers

JAMES CANTON

HarperOne
An Imprint of HarperCollins*Publishers*

First published in Great Britain in 2020 by Canongate Books Ltd.

FIRST HARPERCOLLINS EDITION PUBLISHED IN 2021

Design adapted from the Canongate edition designed by Palimpsest Book Production Ltd

Library of Congress Cataloging-in-Publication Data has been applied for.

ISBN 978-0-06-303794-6

21 22 23 24 25 LSC 10 9 8 7 6 5 4 3 2 1

To Eva, Molly and Joe

CONTENTS

THE OAK PAPERS

BEGINNINGS

S ome five years ago, I sought solace from the ways of
the world by stepping into the embrace of an ancient
oak tree. It is a venerable oak tree, eight hundred years old,
living on the edge of a wood on a small country estate a
few miles from my house. From the first meeting, there
grew a strange sense of attachment I did not consciously
recognise until I later began to realise the significance that
trees, and oak trees especially, can have in our lives. To
begin with, I went there for the gentle comfort sitting
beside that grand oak offered. I could walk away from
my work as a teacher, from my life and responsibilities,
and place myself in a world that was something close
to Eden. I could go beyond my world into that of the
oak. I felt calm descend. Once there I wanted only to
watch the comings and goings of the birds, the bees and
the rest of the creatures that formed the ecosystem that
existed around and about and within that ancient oak
tree. I felt a peace envelop me every time I stepped onto
the country estate where the oak tree lived. Over the
months that followed, I began to visit the oak rather as
one might visit a friend. I became better acquainted with

it. I got to know the distinctive aspects of the tree and the creatures which lived within its realm. I sat beside the oak tree in all weathers and all seasons, at all times of day and night, until I knew that oak like a member of my family.

With the passing of the years, I can now look back and see what happened to me. There was another reason why I sought the embrace of the canopy of the oak. At the time my relationship with my long-time partner had fractured and begun to break down. She and I had started to live separate lives. From somewhere the notion of spending time beside the Honywood Oak came over me. I cannot say now if it was derived from a desire to avoid home or from a need to find some solitude. In truth, it was no doubt partly both.

Wherever oak trees grow around the globe, people have developed a connection to them. Throughout human history particular oaks have been favoured – for their setting, for their age and size. Ancient oaks have always been special. People collect beneath their boughs. They may gather there as a place of significance within the landscape or merely as somewhere to shelter. Whereas we humans are creatures of movement, oaks are static beings. They do not shift. They are born and they die on the same patch of earth. It is that surefootedness that is so appealing. Ancient oaks hold a powerful sense of longevity. The sense of security, the sense of attachment to a place across time, enchants us. We are drawn to old oaks. You can stand beneath a grand oak and know that your more distant ancestors did so, too. Oaks hold onto the memories of earlier generations. By touching the skin of the oak it is possible to feel some tentative trace of those that have gone before.

Human beings and oaks have lived beside one another as neighbours since the earliest times and we continue to do so. We no longer need the bodies of oak trees to build our homes, or to fuel our fires, and we no longer need acorns to sustain us through hard years and meagre harvests. Yet on some level we still lean on oak trees. In ways we do not fully understand, we need them.

PART I

SEEING THE OAK

The Honywood Oak lives on the Marks Hall Estate in north Essex. The tree stands in its own circular enclosure: a low wooden railing that separates the ancient oak from the pine trees surrounding it. The oak has lived here for over eight centuries. Its trunk is knurled and ribbed and close to thirty feet around. The spread of its green-leaf canopy stretches a hundred feet into the spring sky. The tree was a mere sapling when the Magna Carta was signed, when King John reigned over England. As a four-hundred-year-old, its canopy sheltered soldiers during the English Civil War. The owner of the Marks Hall Estate was then Sir Thomas Honywood, a Parliamentarian leader who served in the siege of the local town of Colchester in 1648 and after whom, in more recent years, this grand old oak has been named.

Once there were many oaks, hundreds of ancient trees huddled across these lands. Now there is one. It is a single, lone figure, born on this very foot of soil eight hundred years ago and rose from an acorn to a great tree in the blink of time's eye. I came one day to see this oak, born so long ago, so far beyond the memory of any living human being. I sat in its presence and knew that here was peace.

Beside the coach house, where I parked my car each visit, there is a simple wooden gate that opens onto the heart of the estate. In the moment of stepping through that doorway something truly magical occurred, some form of transformation that allowed me to cast aside all the cares that had gathered upon my shoulders. Once beyond the gate, I ventured into another world.

The gate leads directly into an orchard through which I would walk down into the gentle valley of the stream that weaves through the estate. A path winds to a stone bridge beside a lake. Beyond is the Honywood Oak. The ritual of that short walk was something like stepping back into paradise. Through the gate, into the orchard, down to the lake, over the bridge, up to the ancient oak. The journey took only a few moments. But in that short time, I was relieved of all burdens.

21 June

I am greeted by Jonathan Jukes. He is a calm, modest man. His job title is curator of trees. We walk down from the coach house and across the stone bridge. I look over to the oak as Jonathan starts to tell me the tale of the three hundred other ancient oaks that also used to live on these lands and once formed part of an extensive deer park. In the 1950s, almost the entire population of those oaks was felled for the value of their timber. Four younger trees of three hundred years or so that grew on the garden edge of the gamekeeper's cottage were spared. They live on. The burnt remnant of another aged oak tree known as the Screaming Oak stands crippled and disfigured, yet it has somehow managed to keep life in its boughs. Only one of the truly ancient oaks was shown mercy: the Honywood Oak. It is the sole survivor to remain intact.

The Honywood Oak sits on the border of what once would have been some 2,500 acres of ancient woodland. Quite why this individual tree was spared the axe and the saw is a mystery. The man who has thought longest on the matter, Jonathan – who now acts as guardian to this gracefully ageing tree – believes that the tree must have held some special significance to Thomas Phillips Price, then owner of the estate.

'It may well have been that Phillips Price enjoyed the sight of the grand canopy from the top floor window of the big house,' Jonathan says. 'There's a photograph from that time of a bench tucked against the trunk of the oak.'

He may well have liked to sit under the umbrella of oak leaf that unfurled each spring, to pass a moment in the cool shade of the tree away from the heat of the summer sun. Thomas Phillips Price died in 1932. The big house was demolished in 1950 after falling into disrepair. The truth is that no one alive knows exactly why this one single ancient tree survived the cull of the three hundred other ancient oaks that had lived happily in this hidden corner of England for many hundreds of years.

It is a clear, blue-skied summer's day. The pine trees that surround us were planted as a replacement for the oaks. In that short time, they have risen to the height of the Honywood Oak and now dominate the landscape. We wander through the undergrowth of this young pine plantation where a tribe of pigs had been let loose to eat away the stubborn mesh of bramble which had built up over the years. The pigs have vacuumed the land clean. Eighteen months since they have cleared the ground, a newfound resurgence of life has emerged. Glorious pink foxgloves reach from the soil; seeds that were dormant for years have now sprung.

Jonathan tells of the time when gangs of woodmen arrived in the woods back in the 1950s. A local firm named Mann's carried out the work, which must have taken weeks, or even months. I try to imagine the scene. The chopping and cutting with long-bodied axes and with stretched two-man saws that wove back and forth into the deep flesh of the trees, and so into the heart of the woods.

Robert Burton used the somewhat proverbial line, 'an old oak is not felled at a blow' in his *The Anatomy of Melancholy* (1621).[1] Some centuries later, Thomas Hardy's poem 'Throwing a Tree' pictured the felling of a tree by 'two executioners' who are 'bearing two axes with heavy heads shining and wide / And a long limp two-handled saw toothed for cutting.' In the final line, Hardy writes of 'two hundred years' steady growth . . . ended in less than two hours'.[2]

So how long did it take to fell each eight-hundred-year-old oak tree?

'I met someone recently, at a funeral,' Jonathan says, 'who had been here in those days, who had worked at the task of felling the ancient trees. There may well be others in the surrounding villages who were here sixty years ago as young men, earning a few shillings to turn trees to timber.'

If there are, they will all be old men now.

Ronald Blythe told me not long ago that the ancient oaks are loved by country people precisely because they allow a physical way to commune with and to remember their own ancestors. They are often to be found in the centre of village life, either on the village green or on the well-worn pathways that weave around the hedges and fields. Those trees act as conduits to connect us with those who have gone before. Beloved parents, aunts, uncles, grandparents of earlier generations have also touched the ribbed bark of the tree where now living hands brush the same coarse skin. There is a sense of corporeal connection and so a powerful feeling of remembrance in the simple act of touching an old oak. It is a bodily remembrance through time.

'They had a raw deal,' Jonathan says as we walk in flickering shade.

He is talking now of the oaks that were felled here. He talks of their lives having been taken. We have arrived back at the Honywood Oak. The tree has reached old age. It is 'growing back down', says Jonathan. Just like those last surviving woodmen who swung their axes here in these woods sixty years before, the oak tree is now settling into a peaceful third age.

'They weren't all as old as the Honywood Oak, were they?' I ask.

'No,' says Jonathan. 'I don't think so. Though there were some of considerable age and certainly some were bigger trees.'

'So even older than the Honywood Oak?' I ask.

Jonathan believes so.

There is an inventory of all the significant oaks that once lived on these lands, he tells me, which is held in the Marks Hall archive. After Thomas Phillips Price bought the estate in 1897, his land agent measured all the larger oaks – their height and girth is recorded in ledgers that can still be seen today.

'So you get an idea of the size and the stature of the oaks that were in the deer park at that time.'

I look about us at the tall frames of the pines. I try to see ancient oaks. It is hard to do, hard to imagine the scale of the loss.

'They should never have been cut down,' Jonathan states. 'It is written in Thomas Phillips Price's will that "the oaks must never be felled".'

Yet they were.

One hundred years ago, the ancient oaks of the Marks Hall Estate were not individually named but were collectively known as the Honywood Oaks. At some point after those Honywood Oaks of the deer park were culled in the 1950s, the name became singular.

I am struck with a genuine sense of sadness at their loss.

Ancient oaks have always held our affection but often, too, our awe. Francis Kilvert, writing in 1876 on the ancient oaks of Moccas Park, Herefordshire, knew such contrasting feelings:

> I fear those grey, old men of Moccas, those grey, gnarled, low-browed, knock-kneed, bowed, bent, huge, strange, long-armed, deformed, hunchbacked, mis-shapen oak men that stand waiting and watching century after century . . . No human hand set those oaks. They are 'the trees which the Lord hath planted'.[3]

The trees become 'oak men' and strange and fearful figures. Perhaps in such a transformation comes their salvation. And they are divine, too. No one wishes to cut down and kill revered beings sown by God's own hand.

To lose any single ancient tree is to lose an entire woodland community. Each aged oak is home to a complex, multitudinous collection of bats, insects, birds, fungi and plants. Jonathan compares each ancient oak to a block of flats. William Cowper in his 'Yardley Oak' (1809) saw the tree as 'a cave for owls to roost in'.[4] But it's not merely owls given homes. From treecreepers to hornets, the aged oak plays host to a range of bio-diversity unparalleled by any other tree.[5]

I stand in the wood and strive to imagine the vast lost population of wildlife that would be here were those hundreds of ancient oaks still alive within these soils. Those are the ghost worlds that haunt these woodlands.

3 July

I meet Jonathan in gentle summer rain. We step over the low protective wooden rail and talk about the life of the oak tree. Jonathan says there has been much written and said recently about sudden oak death (*Phytophthora ramorum*). He is concerned for the health of the Honywood Oak. Already other trees across the estate have had black bleeding patches, the lesions symptomatic of the disease. One older tree near the entrance is on its last legs. Others, younger oaks, have suffered the disease and gradually recovered. No one knows how the disease might spread over the next few years. If the Honywood Oak became infected, it could be in real trouble.

'In a couple of years it might be dead,' Jonathan says.

We talk a few moments more in the shelter of the oak. When Jonathan leaves, I watch his truck heading away west through the pines.

I stand and stare. It is the first time I have been here alone beside the tree. For some moments, that is all I do. I am frozen static. Then my hand reaches out to touch the skin of the oak. I brush my fingertips, my palms against the rough bark. Something happens. I feel an ease that has not been felt within me for an age. My heart slows. My spirit stills. I glance around to see if anyone else is around. To be there in the circle of the wooden-railed enclosure, by the oak, suddenly feels strangely like an act of intrusion, an act of trespass into a forbidden space. I am deeply aware of the presence of the oak beside me, but also aware and concerned that someone may appear around the corner of the path through the pines. No one comes. Tentatively, I close my eyes.

Time passes.

A calm creeps over me as though a blanket has been wrapped around my shoulders.

A numinous peace descends.

When I open them, there is only the oak framed before me, the grey bark ridged and still, so still. I feel bewitched. I blink and the oak now seems like a vast and languorous mammoth standing before me with an imposing, giant grace. The bark is bulbous with folds of elephantine skin and the great north-facing coppiced branch is a trunk with vast limbs planted firmly into the earth.

A hare, one of last year's leverets, appears on the path beside the lake, all ears, lean and jittery, and trots gently up the slope towards me only to disappear into the long grasses and ferns that cover the soft slopes of the stream valley. It is a slice of time laid bare, a moment when the normal flow of life is frozen, when instead another sense of being seems to seize my presence in the world and to take a hold over all.

Buddhists call such fleeting moments *satori* – when the facade of our normal existence falls and we see beyond, feel the possibility of enlightenment. We live each day of our life feeling, if nominally, in control of our daily journey through life and knowing our regular path in the world. Then, in such moments of wonder, we can only stand and stare. We can no longer see the everyday. We can only feel our presence here as light as air, our feather-like existence upon this earth as ethereal and fragile as a seed head in the wind.

10 July

The peace, the still presence of the oak, draws me back. On the bench a pale blue damselfly drifts past. It is a tranquil summer's day. I gaze at the tree and think of the oak as a vast compass with each pollarded limb as a pointer marking the positions of celestial objects in the night sky on key moments in the year: sunrise on Midsummer's Day; the transit of Venus; the radiant point of the Perseid meteor shower.

One limb that acts as a pointer to the North Star has been severed, truncated many years ago. A circle of darkness is etched into the round end-face of the pale wood. It is a woodpecker's nest and a single under-feather is stuck to the entrance hole. A spatter of rain falls. Sunshine forces through the cloud and strikes the pines, enriching their bark and greening the leaves on the oak, lighting the patches of bark, the exposed cambium shading, splattering chiaroscuro patterns. On the south side of the oak, what looks at first like a swarm of mosquitoes some fifteen feet up is a bees' nest. Through binoculars I can make out the hexagonal pattern of the honeycomb and think instantly of its close cousin the quincunx lozenge, that blueprint of intelligent design that Thomas Browne in *The Garden of Cyrus* (1658) would have us believe is evidence for the existence of God.

17 July

A muggy summer's day. I feel decidedly delighted to be here, sat on the bench beside the oak. All else in the world is irrelevant. Only here and now exist. My heart leaps a little as I settle on the bare wood. I close my eyes. All is well. All is good.

I look up and see a bramble is growing from a seat in the oak tree some fifteen feet from the ground. On the bench I am hounded, badgered and dive-bombed by a fly which is red-eyed and decidedly devilish. It finally settles to allow closer inspection. I have no idea what it is but I have reference books with me. Could it be some kind of stiletto fly? Perhaps it is even a Forest Silver-stiletto (*Pandivirilia melaleuca*), a fly which is renowned for its aggressive temperament and lives deep in the oak, in the crumbling heartwood core of the ancient tree – the red rot. I only know for certain that this fly now rests on the arm of the bench, with vast red orbs for eyes sticking out from its head, looking utterly alien in mine.

There are speckled wood butterflies flittering through the meadow grasses like burnt ash falling from a bonfire. The air is softly melodious. It is the first sunny day for weeks and all are revelling in the heat, the light. In the south the wild bees are buzzing in and out of their nest excitedly. I step around the circle of the tree. Midges flit in the damp hollows of the buttress where the sun cannot reach. By the truncated pale branch that points to the north, a hornet loops from the canopy, distinct and vast against the azure sky, before vanishing in a single perfect parabola back into the tree. On the bench rests a pile of books on oaks. I have brought a small library yet instead gaze at the life that flies, darts and buzzes about the oak. After an age of damp, unseasonal weather, the world warms.

Oak trees have been on the earth since the end of the Tertiary period some 2.6 million years ago. They have grown in Britain for over a million years. As glacial conditions swept in from the north, so the oak line fell away south; over time as the land warmed and the ice-sheet melted, the oaks returned. The last return of the oaks was around 10,000 years ago, at a time when the eastern edges of these lands were still attached to mainland Europe. As the climate changed and temperatures rose, oaks spread into Britain to colonise those lands freed of ice. Animals came, too, and were followed by humans.

The history of human existence across the northern hemisphere is tightly tied to oaks. Flint axes were forged to fell and split oak trees, along with pine and hazel, cut from the prehistoric forests. These were what are known as the 'wildwoods', which had colonised and eventually covered the land with the end of the last period of glaciation. The hardy souls that lived then used the cut wildwood to feed fires that held the winter from the hearth. The clearing of pine forest by fire and by felling also served to make more space for other oaks to flourish. Around five or six thousand years ago, people started to shift from seasonally nomadic ways to the domestication of animals and the farming of the land. Agriculture began the process of a more systematic destruction of the wildwoods to fashion the land into the first fields for planting crops and to provide grazing for primitive livestock.

Oak trees were special. They gave wood for the fire but their trunks could also be worked to form the frames of the homes that sprung up on the freshly cleared ground. And oaks offered more than merely wood. They gave acorns that could easily be gathered and stored and later eaten so that both humans and animals could feed and stave off starvation in times of famine. There is evidence from prehistoric sites across Europe that those early farmers used acorn flour as a substitute when their own cereal crops failed. At one Bronze Age settlement near Berlin, a metre-deep store-pit was brimful of acorns that had been 'hulled, split and roasted' – that is to say prepared for human consumption rather than as swine-fodder.[6] Throughout human history, acorn flour has been used to make bread. In various communities of the world, the practice still continues.[7]

Those early agriculturists settled into their landscapes and built monuments in stone and in wood. Oak-timbered circles rose from the earth as defined sacred spaces. Tannin from oak bark and galls was used on animal skins to soften the leather. Oak planks were cut and laid out to make tracks that ran for miles through swampy fens and wetlands. The first boats of the Stone Age were canoes cut from single oak logs, while those of the Early Bronze Age were formed of oak planks sewn together with twisted yew branches and made waterproof with beeswax.

Oak charcoal allowed people to progress from Stone Age existence – as charcoal produced by colliers in the oak forests fuelled early furnaces, it raised the heat to such temperatures that their fires could smelt copper and gold and tin from mineral-rich ore stone. The production of bronze was subsequently followed by further oak charcoal-fired furnaces that enabled blacksmiths and swordsmiths to work their magical ways to produce iron and eventually steel.

31 August
The dusk chorus.

Chattering house martins are flitting in broken flight. Echoing about the pines is the incessant *seetoo, seetoo* of great tits. The midges are out. Day is almost gone; only the last threads of light remain. A full moon rises in the east, its face blushed with the colours of sunset. The shreds of night darting to and fro are bats. They weave among the thickening air about the tree. A wren pipes up, a staccato *tic, tic, tic* from somewhere in the darkening mass of leaf in the centre of the oak, clattering away to the cover of the pines while in the brambles below a white angel moth appears bold, brighter in this ever-fading light.

The still that settles upon the oak is that of the closing down of day – a gentle sighing that is broken by the harsh bark of a muntjac deer from the woods behind. The moon, so wide and white, soon vanishes behind cloud. I sit back on the memorial bench, nestling down against the dark. A pipistrelle bat catches in the corner of the torchlight and is framed for a moment, wide-winged. I think of two summers back when I wandered about another wood – Tiger Wood – in the early morning as part of a bat survey, listening to the click of the creatures' echolocation with the aid of a bat detector. The machine told how the patches of shadow that cast about above were those of soprano pipistrelles, the higher frequency of their call distinguishing them from their common cousins. And I was told that night that it had only been in the last ten years or so that the two pipistrelle bat species had been separated by science as distinct. Tonight, the bats flitting across the night are indistinguishable to me. We still know so little.

In the woods behind, the first cry of a tawny owl breaks the night air, two, three hundred yards away. I sit tight and await the nightlife of the oak as the full moon returns through tears in the cloud that appear as though brief glimpses, windows, into another world far beyond. I stand on the planks of the bench and listen to the orchestra of crickets warming up in the bramble, the nettles. Another pipistrelle nips past. The darkness is upon me now. I see little to nothing. It feels deliciously alive to be here, held in this black fog.

Sound is the sense now. The rolling splash of the waterfall beyond is ever present. The crickets tune up. And there are ever wider spaces within the soundscape now the pheasants have lifted into their tree beds and stilled. The birdsong of the day birds is entirely gone. The dusk, too. Now, it is truly night.

Moments pass. A creature stalks the path right beside me and the darkness is such that I cannot tell if it is a fox, a stoat, a deer or a badger. It is gone before I can lift a light towards it. Another owl starts up in the trees to the north. I step down, stand among the trunks, the limbs of trees, and know that here in the forest at night, I have ventured into a forgotten and foreign land; one where ingrained, instinctual senses rise, where fear of the unseen can suddenly leap up, take a hold. The wood after dark is no place for our modern human hearts and minds. Even in ancient times, we went out and away from the woods. We cleared the ground around us in order to live in the spaces created. We carved our homes from the limbs of the trees we cut down. To go back into the woods at night, alone, is to return to another realm, another age entirely. I had expected dusk to fall and the creatures of the night to crawl from under the grand carapace of branches. Instead, there is an otherworldly quiet – a deep, immutable silence that feels like the unheard, slow and steady breathing of the earth. And it enraptures my being to be here. I am so alive. I burn electric. I fizz with frozen energy like a hunter held in the chase.

The midges catch like dust motes in the torchlight. Another creature dances dark-bodied over the pale pathway from the long grasses. I peer. Though the vision is murky, grey, I sense it is a fox. A pinecone falls. I hear it break free forty feet above and there is a moment of pause and then I hear its earth-fall four feet away.

I drink lukewarm tea from a thermos, munch stale biscuits and see that I, too, am a noisy presence, just like the midges and the crickets that break the night's vigil, ephemeral as all others beside this great oak tree. A tawny owl calls to the south. The waterfall falls on the stream. It is hard to rise from this seat where so little and so much happens. I turn back to the oak, so static and silent. Soon I shall have to leave the oak to the dark night, I think. Then a shotgun echoes from the north, piercing the quiet with sudden stultifying violence. The explosion of sound is awful, deathly in its reverberation and it only dies gradually, like ripples across the surface of a tarn, until the tension of the silence that replaces it fades into the same aphonic calm that reigned before.

Walking about the oak in a wide-arced circumambulation, I turn on my torch, seeking any signs of life with the heavy sodium light. The torch beam carves a tunnel of brutal, fierce luminance into the darkness. It is a searchlight in this sea of night, an alabaster flare that shines a

circle of clean light. I came expecting to find birds sheltering, owls awakening. Instead it is slugs that I gasp at –
an olive-patterned, giant-sized colony of yellow slugs that
creep across the western edge of the oak tree's buttress.
There is a Martian quality to them that may be due to the
sudden return of the moon in the sky between bands of
cloud, vaster now as it rises higher in the east, mimicking
the rising of the sun many hours before. And the moon
is now touched with embers of fire and patterned, spliced
with three distinct threads of cloud, transformed from
a pale shadow earth that floats beside ours into another
Jupiter, yet vast and close, burning in the night sky. And
as I stop and stand beside the oak, and lean a hand to
steady me, I wonder if it might be that by some means
unknown I am staring out at some future reflection of
our very own Earth, as the last remnants burn and the
destruction of our planet is finally complete.

18 September

> A forest is in all Mythologies a sacred place, as the
> oaks among the Druids, and the grove of Egeria

Henry David Thoreau, *Journal*, 23 December 1841[8]

When I return to the oak after a period of some three
weeks it is in sunshine and under blue skies. It is late
afternoon and there is the warmth of an Indian summer.

I cannot hold inside the joy that wells within me. Simply to be here – back beside the oak again – is ecstasy. I hurry along the pathway beside the geese. One of the gardeners is mowing the verges around the low perimeter railing of the oak so that as I cross the bridge over the dark lake the air is full with the perfume of cut grass. I raise a hand in greeting. He does the same. We share the same delight.

The Honywood Oak looks magnificent. It stands utterly resplendent, verdant – an arbour of life brimming from the tips of the leaves, the branches reaching forth to hold all within their embrace. On the west side, the chicken-of-the-woods fungus has grown, morphed further with age, and, in having lost something of the vital vibrancy of fresh colouration – of soft, mottled cream touched with breaths of sulphur – has gained an air of permanence, secured in a cleft of the bark some eight feet from the floor. To be within the umbra of the oak, beyond the 'drip-line' – the edge of the overhanging canopy – is to stand under the protective hold of the tree. I step into that realm until I am beside the trunk. A childlike glee overwhelms me. There is a sense of escape from being me, the enchantment of the oak allowing my emotions to soar. All cares and worries fall away when I am here beside the oak. I am happy and glad to be alive again. I have become my own child-self that has fled from the world into the woods. I am alone but I am not lost. When I turn and look upwards through the leaf to the sky my head spins. Sunlight scatters in the oak leaves. A squirrel flies across a top branch like the shadow of a tree sprite.

An hour later, Jonathan Jukes appears beside the pines in his Land Rover. We head over to the west of the estate, to an area that was once ancient woodland. The pines that had been planted in the wake of the culling of the great oaks in the 1950s now rise a hundred feet and more from the ground.

'Over here,' says Jonathan, and forges a path over the bumpy soil.

He halts by a spectral pattern on the earth before him. It is the footprint of a tree. The six-inch raised dais of wood is all that remains of an oak that was felled sixty years or so ago. Like the fossil remains of some vast Jurassic dinosaur, the severed base makes me think of a woolly mammoth's foot. There are five such footprints in this clearing in the pine woods. Each one has been excavated by the action of the pigs, whose patient gnawing away at the undergrowth has scraped free the historical tree record like an archaeological dig. They have steadily exposed the existence of ancient woodland lives that once grew here. They have unearthed a ghost vision of a time when five ancient oaks lived here, each host to their own community of wildlife. The solemn truth falls upon me. We are standing in a graveyard. We remain for a moment in reverent silence.

In *Remaines of Gentilisme and Judaisme* (1686–7), John Aubrey writes that:

When an oake is felling, before it falls it gives a kind
of shreikes or groanes, that may be heard a mile off,
as if it were the genius of the oake lamenting.[9]

That thought of the ancient oaks each crying out as they
are cut down strikes me again as I stand there in the
wood, staring down at the unearthed stumps of the trees.
Those screams have gone. Yet something remains: a silent
remnant, like ashes after a roaring fire. I want to scream,
too, for the lives of those lost oaks.

In the mediaeval world of Dante's *Inferno*, there are
other screams that can be heard in the woods. In Canto
XIII, the trees of the dark wood in which Dante and
Virgil walk are tombs for the souls of suicides. Only when
a scream erupts as Dante breaks a branch of one tree,
does he realise that each tree holds a human spirit. In
the tops of the trees, the Harpies – hideous half-human,
half-bird creatures – make their nests. They sit shrieking
and feeding on the leaves of the trees. As they do so, their
bites tear at the flesh of the trees, causing them to bleed.
Only then can the souls of the suicides within scream
out in anguish as 'the sharp bite gives agony, and a vent
to agony'.[10] In most versions of the tale, the trees are described as thorned, yet in one account I read the Harpies
are said to feed on the oak leaves of the withered trees.

Dante was not the first to imagine such screams in
the woods. Ovid also heard those same cries two thousand years ago. In Book Eight of *Metamorphoses*, there is

another account of 'The Wood of Suicides'. Ovid tells the tale of Erysichthon who commanded his servants to cut down an ancient oak tree in a sacred wood. When they showed reluctance, Erysichthon took the axe and cut down the oak himself. The tree lay bleeding and then it spoke, declaring that Erysichthon would be punished for his violent act, for the oak was sacred to the goddess Ceres. Soon after, Erysichthon suffered from an insatiable hunger and thirst. In order to try to salve his constant craving for food and drink, he sold his daughter and all his possessions. Finally, Erysichthon was driven to tear at his own body and eat his own flesh until he had consumed himself in an especially gory act of suicide. Such were the consequences of cutting down sacred oak trees.

Back at the Honywood Oak, Jonathan talks of 'haloing' the tree – of opening out the space around the oak to provide more light by cutting down some of the pines that are crowding the oak. It is a beautiful word to use, inspiring visions of a luminous circle, an angelic sphere of light. The notion of haloing seems akin to sanctifying the tree with a genuinely spiritual aura. It is nothing new. The ancient Greeks knew of dryads – wood nymphs – the female spirits of the forest trees, who were more specifically of oaks (*drys* meaning oak in Ancient Greek). Ovid's *Metamorphoses* is full of them. Eurydice, wife of Orpheus, was one. Then there were hamadryads. They were also wood nymphs, but far closer to the notion of the living spirit of an individual tree. Trees had a named hamadryad who lived in that particular tree. When the tree died so did the spirit. I look up into the canopy of the oak, seeking faeries, hoping to catch a glimpse of a green-eyed hamadryad. I watch the leaves being gently blown in the breeze from the breath of an unseen entity way up in those lofty heights.

On the west side of the oak, Jonathan points to where there had been bacterial bleeding only two years before. There is a scar in the bark where the infection was, a fissure that looks to have healed over. Trees get ill just as humans do. Older people can die from catching the mildest of bugs. So, too, can old oaks. Jonathan leans in and studies the scar tissue.

Later in the day, I meet up with Jonathan once more. He has walked over from his cottage in the woods. We stand together beside the Honywood Oak. The earth within the enclosure of the wooden railing was hand-weeded yesterday, he says, to ensure safely leaving the oak seedlings that have risen from acorns fallen and buried a year or so ago. They are the next generation.

'You've only got to have one of these to grow,' says Jonathan as he kneels beside one of the oaklets, 'and you've got another one of these in eight hundred years.'

He nudges his head familiarly over his shoulder.

We walk to the south of the oak, look up at the mass of wild bees, then turn to the woodland behind. There will be a little more thinning out done soon; more haloing of the light for the Honywood Oak. Some of the pines will be kept – their evergreen canopies provide protection from the south-westerlies, though lower down their bare trunks offer little resistance against the wind. Other trees will be grown to fill that space.

'We'll plant Lebanese cedars and oaks,' says Jonathan.

And I realise that he is planning not in years but in decades, centuries.

I try to imagine a century on from now. Oaks and cedars and pines form a protective embrace across the south. The Honywood Oak lives on in ever grander maturity.

Across Asia, Europe and North America, the links between early civilisations and oak trees are startling. Where people have flourished and developed forms of established society in these regions, there have been oak trees growing alongside them.[11] Such commonality between oak trees and humans means that wherever oaks grow, they feature in all manner of myths of identity – be they for a defined community, for a nation state or for an entire empire. The oak leaf is a symbol used across the world.

As I have dug deeper into the historical and cultural layers of ancient Europe, I have begun to realise just how common this veneration of oak trees really was.

There are many examples of the ways in which the oak tree forms part of religious practice. In Ancient Greece, for example, the oak tree was associated with Zeus, the father of the gods. At Dodona, in north-western Greece, Zeus's sanctuary was one of the most ancient of all the oracles. In the centre of the temple at Dodona, there grew an ancient oak tree. Here you could come to learn of future truths, whispered to worshippers in the rustling of the oak leaves. That most ancient chronicler of these ages, Homer, recounts in *The Odyssey* a visit by Odysseus to the oracle of Zeus at Dodona where he hears 'the will of Zeus that rustles forth from the god's tall leafy oak: how should he return, after all the years away, to his own green land of Ithaca'.[12]

Those accounts of oak worship in southern Europe in the classical worlds of Greece and later in Rome were paralleled by beliefs and customs in northern Europe, where the Celtic peoples of the same period had their own high priests of the oak called Druids. The very word 'Druid' is a Celtic one, an Indo-European term, formed from the words *dru* meaning 'oak' and *wid*, 'to see or know'. The Druids had 'oak knowledge'.[13] They were the human guardians of the oaks, who held the cultural understanding of the significance of the trees for maintaining the good of society. The Druids not only conducted their various religious rites and ceremonies around oaks; they oversaw the care of the trees and so, too, acted as protectors to the dryads, the spirits of the oaks.

28 September

Two hours after sunrise.

There is a cold, clear light cast by the morning sun though the brushes of autumn are evident. I stand before the oak and sense the thrill of the early morning within me. My breath clouds, condenses into plumes of mist rolling like shrunken sea frets. The oak basks in warm sunshine. Sunlight flickers off all that moves in the canopy – a steady stream of hornets are touched with silver as they emerge from the dark circle of an old woodpecker hole. It is the hornets' time of year. They are imperious insects that tear wasps from honeyed walls of ivy, and that drift through the late summer air patrolling their kingdom.

A blue tit flies above with that splash of blue and yellow that startles. The sight makes me think of the scene in Virginia Woolf's *Orlando* when Orlando holds a yellow crocus and the brilliant blue of a jay's feather and muses on the mixing of those colours. Later, when I return to the moment once more in my copy of the book, I realise I had entirely forgotten that in that same scene Orlando holds the manuscript of her poem 'The Oak Tree', which she began in 1568 and had been working on for close to three hundred years.

White strands of cirrus cloud are creeping in from the west, fracturing the light. A treecreeper is etching a spiralling path about the skin of the central body of the oak. I gaze upwards at the treecreeper as it winds around the vast southern limb heading skywards, twisting about the bark, a flickering, busy passage, coiling higher and higher into the far reaches of the oak. I am my child-self again. Ecstatic in the freedom of being here alone, I delight in the solitude of simply standing in the early morning sun. A breath of wind brings a pitter-patter of last night's raindrops from the uppermost leaves on the highest reaches showering down into the lower canopy, splashing my face. The touch is cool, fresh and delicious. My senses scatter with the glory of the moment. The birds know the same gladness. Blue tits *tsee tsee* endlessly above me in a monotony of glee.

When I walk away it is to follow the path about the great lakes and the two acres of walled garden and when I return to glimpse the oak again I see autumn is here – a touch, a colouring of soft yellow, a handful of darker, copper tones amidst the green. In the woods behind, the scent of pine settles like a moment of still. A family flock of long-tailed tits flit among the jade needles, shooting like darts from tree to tree.

2 October
Two hours till sunset.
A gentle southerly threatens rain from patchy cloud. The day feels done though the dusk is a while away. I sit on the bench and wonder a duality of desires: to care for the oak and to be cared for by the oak. It is a growing feeling that sits insecurely in me. I cannot frame the sense yet for the past few visits to the oak I have increasingly felt an unfathomable notion that the oak is in some way

protecting me. For a while, I think of the delighted child that I am able to become beneath these boughs and yet as I wonder on that fact I know that in doing so I endanger the becoming by destroying the enchantment. To think too much upon the matter is a mistake. So I rise and step over the low hurdle of the wooden-framed enclosure rail, within the drip-line, under the umbrella of the canopy. I gaze upwards and feel again the protective umbra of the oak.

An hour later, I peer across the stream to the oak and notice a few more leaves have turned to paler autumn shades like the grey hairs on a father not seen for months. By the lake, beneath the yew, is another bench. I watch the oak. Another leaf falls. A crow leaves. A jay flies by. It has been another fallow year for acorns from the oak. In the air above the low wooden rail a hawker dragonfly quarters the enclosed space like a barn owl at dusk across a favoured field. Cloud gathers in the south. Rain falls. I shelter beneath the yew. Beside the lake the old grey heron waits in eternal vigil. The cold breath of autumn is there in the late wind that ripples the skin of the water, brushing all with goosebumps. I huddle. On the bridge, a grey wagtail bobs up and down, up and down, a splash of yellow underbelly gloriously vivid in the gloom. Above, in the darkening sky, a pale swan and four cygnets muddy with young brown feathers have lifted from the lake and now soar away through the waves of cold rain that sweep in from the south.

9 October

On the mossy north-west face of the oak, I hunker down.
A ledge a foot or so from the ground is my seat; my back
nestles into a hollow. It is a sacred space, an eremitic
space where I may go to rest and pray and find peace.

A dark-fingered wing flickers beneath the leaf line, landing somewhere behind me, by the severed north-facing
limb. We are on opposite sides of the tree trunk. My body
leans against the bark. A diameter away, the woodpecker
grips tight. I wait. In a few minutes there is one cry, a single yelp. A moment later, she flies. A blush of red under a
black cloak, a shadow across the green leaf, up and away,
rising and falling in flight to the bridge, the lake.

12 October

There is a wedding at the coach house today. Jonathan has
warned me, yet still the noise and bustle, the screaming
children, the happy shouting of drunken adults come as
a shock.

By the oak, it takes time before any kind of peace
settles. From a clean blue sky pine needles fall, rustling
together on the littered floor like tiny feet in dry grass.

The racket from the wedding party is too much. It
drives me away. I follow the stream south. A solitary
jay flies in from the east. A clutch of yellowhammers
sweep overhead, and their urgent, infectious calls fall as
perfectly thin slices of sound with a windblown scatter
of leaves. Autumn is here, they cry. Winter is coming.

A week ago, I had a delivery of logs for the wood burner. The local woodman Richard Fordham brought them himself rather than one of his lads. In the dying light, we pawed handfuls of cut wood from the back of his battered red Toyota truck, the logs clattering down onto the concrete driveway. I told him of the Honywood Oak. Richard knows more about trees than almost any man. He has worked with them since he was a boy. He listened as I told the story of the felling of the ancient oaks.

'You want to speak to Richard Cromach,' he pronounced. 'He worked at Mann's. Taught me from a kid about trees.'

He paused.

'Like a father he is.'

Richard looked up from the stack of wood.

'You want his number? He'll tell you about the oaks.'

In the shed, I searched for pen and paper and returned with a stub of pencil and an aged Penguin paperback. Richard relayed the phone number, which I wrote on the back of the book.

'I love trees, I do,' he said with wide eyes gleaming through the gloom. 'There's one down in Huntingdon. Willow, it is. Most amazing tree. Huge.'

As he went on I could not help but think that here before me really was a Green Man, or indeed a modern day version. Not carved and hidden in a church nave but alive and sharing with me his love of trees, his years, his decades of wood knowledge.

16 October

Tucked beside the oak, huddled against the bark in the calm of the west face, a still starts to fall across the world. Now is the time of year when daily a few minutes of light are lost. In a fortnight from now, an hour of light will have vanished from each day.

The sun sets earlier and earlier. We do not plunge into darkness. The days fade. The darkness of winter settles upon the earth slowly like a vast cloud that gradually floats across the sun and steals the light.

Nestled in the embrace of the Honywood Oak, a silence falls. At first it seeps gently into existence, as the voices of others fade. For a time, there are distant if distinct noises: of the birds that continue to call, the geese, the blackbirds in the ivy, the long-tailed tits in the fir trees. Then those voices also fade. Nature is often silent. It does not always speak to us. In winter, that speechlessness is ever more evident. Think of the intense muted hush of snowfall. That is the truest silence of winter.

I ring Richard Cromach and ask him about that time sixty years ago when there were still hundreds of ancient oak trees on the lands surrounding the Honywood Oak. He tells me what he remembers. It was another age entirely. He was a young man then, he says.

A week on, Richard is standing outside the Co-op on the High Street in Earls Colne. He is dressed in a smart tweed jacket and has a box of remembrance poppies tied around his neck. He is there every year. I am passing with my daughter Molly but stop and joke with Richard as to how long he intends to stay there. It's freezing cold already and only halfway through the afternoon.

'Not much longer,' he says and smiles and wipes a drop from the tip of his nose. I can see the same glint in his eye that I'd seen in other woodmen, something deeply resonant of the time they'd spent in the woods, with trees.

Some kind soul from the shop over the road has given him a cup of tea. We start to talk again of his time working in the woods, of his time over on the Marks Hall Estate. There's a question on the felling of those ancient oaks that I've been wondering about for a while.

'How did they cut the oaks down back then?'

I have this image of huge, toothed saw blades being drawn back and forth by hand through the body of each vast oak. I imagine a forest stilled but for the monotonous sound of the saw cutting away steadily at the heart of the tree. I am mistaken. I have suspected it for a while. Richard dabs his nose once more with his handkerchief.

'Were there chainsaws back then?' I ask and feel like a child. My own child Molly moves beside me, her hand reaching down to stroke a passing dog.

'They did have chainsaws,' says Richard. 'Not like

today's though. Great big things. Took two men to work them.'

And I try to imagine the scene in the woods up at Marks Hall Estate as the ancient oaks are felled. Gone is my sense of two-man saws and a steady, silent intent; of fierce axes thumping into the oak wood. Now I hear only a violent noise and roaring machines tearing away at the oaks, the primitive chainsaws as they snarl and growl and bite their way through the bark, on into the flesh of the oak, eventually severing them from their roots. I remember those vast dinosaur-like footprints on the woodland floor that Jonathan Jukes had shown me – the stumps of ancient oaks that had been cut down.

I buy a poppy and pin it to my pea coat and thank Richard and wish him all the best. Then I walk on up the road with an arm over Molly's shoulders. When she scoots on ahead I am left wondering upon those times sixty years back, thinking of this kind man standing in the cold, then a teenager, a kid, in the oak woods.

6 November

Three weeks on, I return to the embrace of the oak, tuck myself once more into that cranny in the western face, my back against the oak tree's bark. The dusk comes early now. The day is grey; the cloud hangs low. By four thirty, the dark will be here.

As I arrive today and walk round the yew hedge it is to meet a gang of greylag geese who stare and wander away from the slope of the stream. The rain has settled in. The day began with sunshine, a sun that rose a street-light orange to clear skies, promising hope. Now rain – sodden rain – soaks all.

When I glance from the geese to the oak, something has shifted in scale. The tree looms larger, wider. The bank of pine trees to the west, planted soon after the massacre of the other oaks in the 1950s, has been cut down. From a distance of some two hundred yards, the circular end trunks, gathered in timber stacks, shine with a honeydew hue. Beyond the bridge across the lake, the damp air is drenched with the sweet perfume of the pine, cut a few days before, still pungent, even in the rain. Up close the mandala faces of the pine trunks hold an almost perfect radiance. Circles on circles of growth rings telling of a half-century lifetime reaching for the sky with vertical precision. The rings of years are easily countable.

The space is shocking. Where there was once wood, there is now air. Here is more gentle management: the considerate and careful healthcare of the oak. A shelter belt. I had talked through the term with Jonathan some weeks back. The winter winds, the storm clouds, would be at their worst from the south and the north so he would leave the protective gathering of those pine trees. To the west, the taking down of the pines would open up the oak to more sun, more light. 'Haloing': that was the word. In spring, that stretching space, that legroom, will be felt by the Honywood Oak. The widened arc of sky will be appreciated. The first stirrings of new shoots on those exposed western limbs will signal new growth from the old tree.

The light of the day has gone. With the changing of the clocks, the throwing back of time, the dusk appears more suddenly, intruding into the precious daytime. As the light thickens, I tuck into the embrace of the oak and peer to the newly framed horizon. The felling of the pines has flooded the west with air and light. At my feet the eviscerated shreds of those pines have been scattered as mulch in a dark, humus-rich circle around the oak, the circumference formed by the drip-line. It will nourish deep roots; keep the weeds at bay.

17 November

Once more it is only moments from dusk. A cold southern breeze drifts over the stream. In the open corridors of air where the conifers were, pipistrelle bats now fly, floating in broken circles, darkening the day even further.

Autumn closes in. Unseen shady holes in the upper limbs of the oak become visible as the leaf canopy falls. The woody mulch that rings the oak is now coated with a layer of leaf that shines copper even in the fading light. A grey squirrel dashes across the blanket of oak leaf on a path that leads straight to me. It skids to a standstill, scattering fragments of wood, tearing dark claw marks in the floor. It stops utterly still; then scarpers, away to the east side of the oak, appearing a moment later ten feet above, raining bark litter down on my head. Another emerges, sprints the radius of the railed enclosure then flees as nine geese fly overhead, a perfectly framed and silent skein – dark against the grey light. Colour fades. My eyes start to see with rod cells. Though they allow good sharpness of sight, they see no colours. It will take two hours of darkness before the shift from the light-sensitive cone cells is complete: to night vision. Clear and colourless – like moonlight.

30 November

By the time I reach the oak, the sun has already risen to reveal a cold, clear day. Frost lies on the ground. In the west, the moon is full and pale. Jupiter sits beside it – a glimmering white bead. All augurs well.

Simon is already there. A ladder rests against the south of the oak, leaning some twenty feet high. He is busy looping a length of red rope around a distant branch. We shake hands, share a few words on this glorious morning.

'You done any rope climbing before?' Simon asks.

'Not much,' I admit.

We walk back to the Land Rover. He asks exactly how much.

'None.'

My plan had been that I would climb up into the heart of the oak to spend time just being there: to see the comings and goings of insects, birds, animals; and something more – to simply be. I had hoped to spend some time up within the embrace of the oak; as I now commonly sat at ground level in that nook of sacred space on the west side of the oak. I thought I would merely sit there some hours. To observe and to feel.

Now, the plans have changed.

'You have to be harnessed,' Simon says. 'And I'm afraid I'm not allowed to leave you up there alone.'

Such are the rules of the insurance on the estate. It is a matter of health and safety. My concerns had been rather different. There is a sense of trepidation but one born of a disquiet at climbing up into the tree. It seems like an overstepping. I want some kind of invitation; some sort of sign that I am welcome to venture further up into the body of the oak.

By the ladder, I wait. Simon works away at securing the rope system so that we might climb safely about twenty feet up in the air. At the foot of the tree, he picks up something from the ground.

'Honeycomb,' he says.

I know exactly where it has come from: the dark hole above our heads that is an entrance way for the colony of wild bees. I lift the fragment from the leaf litter. The structure of hexagonal cells has an impossibly perfect feel. How could bees have created this? I am suddenly struck by the brilliant beauty of the shape. I lift the section to my nose and it yields an incredible smell: a dark, rich, sweet honey aroma that is held within a deeper, earthy layering similar to the pungency of fresh fungi. I place the honeycomb carefully on a seat of bare wood at the foot of the oak. It is only later that I realise it was the sign I had been after.

As I step up into the oak it seems as if I reach over a threshold of some kind. To lift from the earth, to climb into the tree, is to enter into another realm, another landscape. To ascend from the ground is to rise above the world. So it feels as though it is a sacred step that takes me from the frame of the ladder across and into the oak.

'Just lean back into the harness,' says Simon.

I lean back. His rope tricks ensure that we can lurch and lean from the branches with the secure abandon of apes, our bodies held above the ground by a sophistry of swings. In order to get to our present state of comfort, we have moved with a certain apprehension past the dark circle from where the honeycomb has fallen.

'You don't want to wake the hornets,' one of the gamekeepers calls up helpfully, grinning with delight at the thought. But the hornets sleep on. In reality, they are bees anyway. The honeycomb offering tells the truth of that.

At twenty feet up, the familiar landscape about the oak looks transformed. Our vantage and the enhanced perspective make the shelter belt of the conifer woodland far more solid, corporeal. Seen from the ground the pines are more space than wood, mere stanchions, stems. From a height they bulk out into greener masses. Scanning the scene, the lake seems larger and the stream more sinuous as it weaves south through the estate grounds. But it is the oak itself that is transformed more than anything else. The main branches, those limbs reaching to the cardinal points, are absolutely vast when viewed up close. Simon wraps his arms around the south-westerly facing branch, failing to get anywhere near touching fingertips.

'Just this branch is like a two-hundred-year-old tree on its own,' he says.

I have been reading up on Druids and Green Men. Simon is very much a modern version, not that he would have framed himself in such terms. Notions of long-haired wildmen of the woods don't fit well with this modern woodman. He is what is now known as an arbo-riculturist. He knows his trees, works closely with them, cutting and shaping, managing their growth. Yet there is also a deep-seated respect, that same thread of some hard-to-describe tree spirituality which ties him to his Druidic ancestors.

'I love the moment when you're alone in the tree,' he confesses. 'High above the ground.'

He has never been up into the Honywood Oak before. Today is a pleasant excuse to leave the other trees for a day. We hang from our rope lianas and chat.

'Got to cut some mistletoe from those limes, some-time soon.'

He points over to the far east of the estate, beyond the lake, where an avenue of tall, lithe trees hold the stringy balls of mistletoe high in the air.

'They move about a bit,' he says with a smile.

We clamber about the limbs of the oak like oversized treecreepers, then into the heart of the oak – the space which has formed at the head of the main trunk over the previous centuries as the arterial branches have grown out leaving a clear open bowl, a palm. From below I have always imagined climbing up into that world, tucking down there and passing the day, the night, feeling closer than ever to the oak. Now, I am there. From within the crevices of the severed facade of a main limb, a cluster of fungi grows: dark brown caps over white stems. A gnawed pinecone rests on the smooth surface of pale oak wood where the bark has been stripped free. This lofty space is even more enticing. I want to settle in here and remain. Here is a whole world of its own – another inscape within the oak. I search for owl pellets but find none. In the rotten hole where a branch once grew, the copper litter of leaves hides a foot-wide clump of cement used some time back as crude filler.

'Never do that now,' says Simon when I relay my finding. Arboricultural practice has changed.

I peer gently into crannies, at the aquamarine, the lime-emerald tints and hues of mosses and lichen so vivid. Sunlight glances, glitters off the shades of leaves turning brown from green. Above, the twisting branches flare and fly into the blue sky like witches' hair. My head spins a little. I look down to the carpet of copper leaf that shines bronze, to the shaded sections hidden from the winter sun by the oak, still whitened with frost. It is a turning time of the year. An autumnal fare of auburn patched with the green of leaves still living.

When we step back to earth there is one task that remains. Simon heads away to the store sheds leaving me alone again by the oak. A blurred flutter of black and white dives across the space of the west shelter belt – a greater-spotted woodpecker that pauses on the trunk of a pine. I creep across the rolls of cut pine until close enough to glance at the glorious scarlet of his tail. From the cold shadowlands of the shelter belt, the ladder leans incongruous against the oak. Simon returns in the Land Rover with a measuring tape. In a final ritual, we wind the tape round the girth of the oak – like May Day ribbon – at a height of four feet from the forest floor.

'Twenty-eight and a half feet,' declares Simon.

This translates to an age of some eight hundred years. Just as we thought.

PART II

KNOWING THE OAK

At first, being beside the Honywood Oak gave me the space to simply be in the natural world and to think. It is hard to find the chance to step away from the frenetic busy-ness of modern life. In the peaceful enclave of the country estate it was easy to pay attention instead to the ways of nature, especially when beside a presence so huge and yet so silent as the oak. At first I was ecstatic in my delight. I thrilled at learning of the ways of the oak, its ecosystem and all the collective life of the creatures that lived within the enveloping arch of that ancient tree through the shifting seasons. I was full of natural wonder. In notebooks, I began to keep diary records of each visit, each moment passed beside the oak. I started to gather together lines from poems that I came across, from essays and novels and histories and works of philosophy that also spoke on oaks and of their meanings to us, their significance in our worlds. I came to know the ways of the Honywood Oak. I now realise that in a sense I fell under the spell of the oak.

I also learnt of other great oak trees that grew in these

lands and others further afield. Each had lived for hundreds of years and meant much to generations of human beings who had come to stand beneath their boughs through the centuries. I found aged and glorious tomes of books that contained intricate portraits of ancient oak trees; some that still thrived, some that had since fallen and gone but remained in perpetuity through the hands of those patient artists. My oak diaries swelled with entries. I learnt of the flora and fauna that lived beside and within the oak. I read and wrote of that myriad of plants and creatures and then saw and experienced them about the ancient oak. And as the pages of my notebooks filled, I knew that my growing wonder in the ways of the oak was mirrored in the minds of so many others that had been before me. That knowledge made me feel ever more grateful, the more I came to know.

I turn to Pliny the Elder, the Roman author who is the most ancient of writers to offer some written comment on the ways of the Celts in Britain and their reverence for oak trees. Pliny began his encyclopaedic *Historia Naturalis* in 77 CE, two years before he was killed in the volcanic eruption of Mount Vesuvius. In this vast tome on all matters of the world, a window is opened into Celtic oak practices:

The Druids – for that is what they call their magi-
cians – hold nothing more sacred than mistletoe and
the tree on which it is growing, provided it is . . . oak.

It was the oak that inspired greatest awe in the Druids.
All possible elaborate care and attention was given to
the ceremony. Dressed in a white robe, the Arch-Druid
would climb into an oak and cut the mistletoe from the
tree with a golden sickle, allowing the sacred plant and
its berries to fall and gather in a white cloak held beneath
the tree. The mistletoe not only served to fertilise any
barren animal but would also act as an antidote to all
known poisons.

Pliny told how the gathering of the mistletoe from
the sacred oaks was undertaken 'on the fifth day of
the moon' and how a sacrifice of two white bulls was
made at the same ceremony once the mistletoe had been
gathered.[1]

It was easy enough to close my eyes and see that pale
sheet under a vast oak and to catch a glimpse of ghostly
moonlight on the golden blade that cut the mistletoe
from the highest boughs.

Pliny's words found a wide new audience in Victorian
times when his depiction of the Druids and their oak
worship was revisited by the pioneering anthropologist
Sir James Frazer, who was quick to clear up the matter
of the ancient British priests' name. Pliny was confused
on the etymology:

Pliny derives the name Druid from the Greek *dru-s*, 'oak'. He did not know that the Celtic word of oak was the same (*daur*), and that therefore Druid, in the sense of priest of the oak, was genuine Celtic, not borrowed from the Greek.[2]

Sir James Frazer's *The Golden Bough* also recounts how other prehistoric peoples across Europe also practised 'tree-worship' and often paid particular worship to the oak tree.

Frazer tells the tale of the Boeotians of Plataea in Ancient Greece who held a festival known as Little Daedala. He explains that all the Boeotians would head out to 'an ancient oak forest' where the trees were of 'gigantic girth'. Once in the woods, they would place 'some boiled meat' in a clearing and then stand back and watch as the birds came in to feed on the offerings. When a raven was seen carrying off a piece of the meat, the bird was followed to whichever oak the raven settled upon. That tree was then cut down.

The wood of that chosen tree was then used to carve a figure that was dressed as a bride and put 'on a bullock-cart with a bridesmaid beside it' before the cart was drawn off to the banks of the River Asopus and then back to town 'attended by a piping and dancing crowd.'[3] These oak figures were then carefully stored until the celebration of the Great Daedala which occurred only once in every sixty years. All the oak figures gathered from Little Daedala celebrations were then taken to the summit of Mount Cithaeron and gathered at a wooden altar specially constructed for the event. Sacrificial animals and the entire collection of sacred oak icons were then set aflame in a conflagration that could be seen for many miles.

The ties between birds and oaks in ancient times are often made. It is the raven that chooses which oak tree is to be sacrificed. The Boeotians are merely human onlookers. When Robert Graves turned to the matter of the oak and the Druids in *The White Goddess* (1948), he added the detail that 'the Wren, *Drui-én*, is the bird of the Druids . . . And the wren is the soul of the Oak.'[4]

16 December

Hard by, a cottage chimney smokes,
From betwixt two aged oaks.

John Milton, 'L'Allegro' (1645)[5]

It is a Sunday. Watery winter sunshine falls against the oak. Few of the leaves now remain. I sit on the epicormic hump on the east of the tree. A jay saunters past with slumberous flight towards the pines in the north. I watch as motes of minute flies that might be midges spiral in the low sunshine. The balls of mistletoe sit high in the lime trees.

Later, back in the books, I find a suitably seasonal oak comment from John Claudius Loudon's *Arboretum et Fruticetum Britannicum*:

> Baal [the Celtic god of fire] was considered the same as the Roman Saturn, and his festival (that of Yule) was kept at Christmas, which was the time of the Saturnalia. The druids professed to maintain perpetual fire; and once every year all the fires belonging to the people were extinguished, and relighted from the sacred fire of the druids.

According to Loudon, this was the origin of the tradition of the Yule log, whereby a fresh log of wood was placed on the fire and then removed before it had burnt completely such that it could be kept until the following year to light the fire on Christmas Day. So the ancient British Druidic practice of the sacred fire was maintained in the use of the Yule log at Christmas time. The Yule log had to be, of course, of oak.

I stare into my wood burner and muse on the notion of perpetual fire and of the Yule log of oak. I wonder if anyone still burns Yule logs of oak at Christmas. Perhaps we should. Perhaps that is where we are going wrong.

And then I find another little nugget of oak knowledge from Loudon, who declares that:

> The worship of the druids was generally performed under an oak; and a heap of stones was erected, on which the sacred fire was kindled, which was called a cairn . . . from kern, an acorn.[6]

As I have read more and learnt of ancient ways, I have recognised how human beings across many continents, and for the vast majority of our existence, have been living in a world far more closely connected to oak trees than we do today.

From the earliest Stone Ages through to the start of the Industrial Age merely a few hundred years ago, we have been dependent upon oaks and infused with worship of those trees, and religious practices centred on the most esteemed of the oaks that lay within our communities. Our wisest people, our most thoughtful leaders, have been those who served to oversee the links between us as living humans and the deities of the oaks.

These were ways and practices that are now called pagan. Yet I learnt some time ago that the word pagan merely came from the Latin term *pagus*, meaning 'of the countryside'. Whereas 'civilised' peoples were those of the towns or cities, 'pagans' were simply those from the rural worlds beyond the city walls. The gradual shift that has taken place across the world as humans have moved to towns and cities in the last few centuries, has been one inevitably accompanied by a movement away from innate associations with the natural world. In Britain, there existed few towns as we know them before the arrival of the Romans. Instead, the entire country was centred on life in the countryside. The population was, by definition, pagan. The worship of gods and spirits associated with oak trees shifted not only as monotheistic religion won out over the religious ways of the people, it diluted and changed, too, as people gathered together in towns, away from the forests, the land and the oaks.

With Christianity came an inevitable further realignment of the relations between people and oak trees. Yet the relationship between the Christian Church and oak trees was not always clear-cut. Gospel Oaks were those tree-marked places on the edges of the villages where preachers held open-air services to their gathered congregations. In rain they had some shelter and in sunshine they had some shade. And what then of the figure of the Green Man – an image that can be seen carved into the oak beams of the finest churches and cathedrals across Europe? A face formed of a fusion between the features of a human and the leaves and branches of oak trees is actually rather common in Christian sites of worship, especially in certain English churches where the foliate faces of Green Men peer from beneath the elaborate camouflage of intricately carved flora. Many of the Green Men figures to be found can be identified with Sylvanus, a Roman god of fields and forests, and so closely associated with trees and oaks in particular. The foliate head is the form common to all Green Man motifs and is commonly dated from Roman art of the later first century CE.[7]

But the Green Man representations are also reminiscent of a wilder side of human nature that can be found in one incarnation or another in all civilisations and certainly dates back 5,000 years in Mesopotamia. There is the figure of Enkidu told in that most ancient *Epic of Gilgamesh*. I can picture the cuneiform markings in the clay tablets where the tale was first recorded. Enkidu was a wild man, raised by animals and brought into civilised ways gradually. He was perhaps some kind of proto Green Man. Yet the wild man always appears in a more ferocious form than that later embodied as the Green Man across Europe. While the wild man represented the antithesis of the civilised man of the town – Enkidu as contrasted to the urbane Gilgamesh – the Green Man was the personification of the spiritual dimension of the woods.

Even as we moved away from the forests, we kept him in our thoughts. So the Green Man took on various forms. He was Robin Goodfellow, woodland spirit, the Puck of Shakespeare's *A Midsummer Night's Dream*. He was 'The Green Knight' who appeared alongside Sir Gawain in the fourteenth-century Arthurian poem. He was the Old Man of the Woods and Jack-in-the-Green. He was a humanised form of the dryad, the spirit of the Druidic deities that was still seen to dwell in oak trees even in Christian times. And he had corresponding forms, too, as the Green Woman, and as Green Children: a family of oak spirits that continued to live on, both in the forests and trees that grew beside our homes and in our deepest cultural remembrances.

Occasionally, Green Children have actually been witnessed. The thirteenth-century monk Ralph of Cogge-shall gives an account in his *Chronicon Anglicanum* of two children found in a village in Suffolk then called St Mary of the Wolf-Pits (now called Woolpit) after the traps laid thereabouts to catch wild beasts. The children were discovered one day beside one of these pits. They spoke in another tongue and their skin was coloured green. The boy was apparently sickly and died soon afterwards, but the girl survived and in time, when she had learnt to speak English, told how they had come from a place where all the people were a green colour.[8]

Then there is the tale of Peter the Wild Boy. When he was found around 1725 in woodlands near Hanover in Germany, the boy was thought to be about twelve and had apparently been abandoned in the woods by his parents many years before and had survived by eating leaves. His feral state was not complicated with green skin. He crawled about naked, spoke no language and captured the interest of writers Jonathan Swift and Daniel Defoe, who wrote of this remarkable creature as 'mere Nature'. King George I had him brought to the English court and when Peter was painted there, it was as some form of Green Child, with oak leaves and acorns in his hand and a green coat around him.[9]

These Green Children seem to haunt me, to run around my head. I know I will have to return to them, to tell more of their strange tales, their extraordinary lives.

4 January

After a season of wet rain, an hour or two of winter sunshine feels heavenly.

The Honywood Oak is almost entirely denuded of leaf now. A gathering of greylag geese squawk and squabble messily in the mud while they graze the green baize beside the lake. Two black-nosed Canada geese step wearily about the more raucous greylag like tired parents. Gleefully, the youthful greylag chatter and chirp.

I sit on the bench. One of the remaining pine trees creaks in the wind behind me like an old staircase in the night. It is a sound that seems to echo the calls of the long-tailed tits which flit unseen in the trees beyond. From the west, a lone brown leaf turns and tumbles through the air. It is an oak leaf. It turns, spiralling in a parabolic descent and lands on the collected carpet of leaf litter from the Honywood Oak.

15 January

Cold. Snow.

A dozen goldfinch perch freezing on a young oak by the site of the old house. I hear their plaintive cries as I walk over to the Honywood Oak.

Snow fell last evening as darkness came. Vast white eggshell slivers of snow tumbled through the air and soon gathered as a pale layer upon the ground. In the lee of the oak, facing west, I sit with my back to the coach house. A vestige of the snow still coats the land around. Drops of melted ice fall from a hidden tap high in the treetop – tainted, bark-tinged droplets the colour of peat. Within the umbrella of the oak, there are stiletto holes in the surface of the snow that mark where the oak has melted the snow upon its branches. For a while, the afternoon sun flares boldly to a glorious orange circle, then hides behind cloud beyond the pine trees. A flock of greylag geese overhead break the still and fly through the departing sun, splintering the light. From the north, snow clouds roll in. Something oddly otherworldly overcomes me, as though I have become caught in another reality, swept up into some strange dream.

John Keats in his poem 'On Sitting Down to Read King Lear Once Again' wrote a plea that 'When through the old oak forest I am gone / Let me not wander in a barren dream'.[10] The notion that the forest induces a dream-state was nothing new. Woods and forests were places of sprites and spirits and fairies and dreams and transformations. Shakespeare had shown that well enough in *A Midsummer Night's Dream*. Yet Keats draws the distinction between the 'barren dream' and the fruitful dream. Both could be induced by the ancient oak forest. One is welcome; one is not.

In *The Fairy Mythology*, his 1828 collection of tales about fairies, gathered from around the world, Thomas

Keightley gives a 'goodly catalogue' of some of the spirits mentioned in an earlier work by Reginald Scot titled *The Discoverie of Witchcraft*. Included in that list is a figure known as 'the Man-in-the-Oak' whom Keightley notes is 'probably Puck' before heading to another source for evidence:

Turn your cloakes, quoth hee, for Pucke is busy in these oakes.[11]

These words are taken from the poem 'Iter Boreale' by Richard Corbet (1582–1635), a bishop who enjoyed writing the odd verse of poetry. Turning your cloak inside out was considered a charm against the actions of fairies. Quite why, I do not know, but the action is remembered in the traditional English folklore rhyme:

Turn your cloak for fairy folks are in old oaks.

The 'Pucke' mentioned is the same Puck who appears in Shakespeare's *A Midsummer Night's Dream*, often going under the name Robin Goodfellow or merely Robin. The likelihood is that Shakespeare had read Reginald Scot's *The Discoverie of Witchcraft*, which had first been published in 1584 and remained popular throughout the sixteenth and seventeenth century. So the Puck who we read of in Shakespeare is likely to be an impish incarnation of Scot's 'Man-in-the-Oak'; that is to say a re-imagining of the oak dryad of ancient Britain and Greek mythology.[12] Some centuries later, when Puck appears in Rudyard Kipling's *Puck of Pook's Hill* (1906), he describes himself as 'the oldest Old Thing in England'.[13]

The fairies were sometimes mischievous, sometimes merely playful. Their presence in the ancient oaks was persistently felt by the country people. They even ventured into the cities so long as there were old oak trees to provide a home. Elsie Innes' *The Elfin Oak of Kensington Gardens* tells how:

The Wee Fairy Folk have come to London Town. Have you seen their home in the old, old oak in Kensington Gardens?[14]

25 January

Dirty white skies mirror the land. Snow that fell through the darkness of Monday morning still carpets the rough ground. I walk towards the oak past the round patches of grass, circles of brown earth that have grown beneath the leafless canopy of trees, formed from melted snow dropping from bare branches.

At the oak, patches of ochre have emerged to the south of the tree revealing strips of oak leaf mosaic. To the north, all is white. The stiletto-holed splatter pattern of melt drip is insufficient to erase the blanket of snow that remains.

The oak is so silent in the snow. Snow suffocates sound. All is still. And then into that void the tiniest noise now echoes loud. I brush snow from the bench to sit a while though the cold is soon bone-chilling. In this deep winter, it is easy to see the oak as having shrunken back somehow. Stripped of leaf and life, it seems to have tucked down, to have been distilled into a central core of being in order to survive, just as a colony of bees dies back to become a single hibernating queen through the dark months. And isn't that, after all, what we do, too? Through winter we tuck back deep inside ourselves until the spring sun draws us out again.

I feel my fingers freezing from the tips, my toes tingling with the cold, and wonder at the static nature of the oak in this frozen land. Humans are not born to stand on the same square of ground. I must move. I rise and walk away to keep the warmth of life flowing through me, stamping feet in the snow and clapping hands and striding now towards the conifer plantations to the north beyond the lake where the land is warmer. An emerald path runs beneath the statues of spruce trees. Snow rests in patches on the evergreen branches and in the brambles that cover the forest floor. To the west a buzzard mews. I halt. Sounds return. I hear a woodpecker drill. Then, as though from some far forgotten realm arise a choir of tiny piped chimes that ring out from above – goldcrest flit and flicker as sprites in the trees. One ventures towards the ground. A few feet from me, its bold golden head-band is as bright against the snow as unearthed ancient treasure. There are two distinct calls: the louder pipe of one, lone, individual bird; then the lighter chime of the rest that follows, which is angel music and spellbinds all.

When I return to the oak, a green woodpecker stands on the snowy ground within the umbrella of oak branches, tossing leaves and snow; seeking something, anything, from the frozen soils.

8 February

A cold northerly drifts down south. Winter sunshine shines through a cirrus sky.

A mistle thrush flies from the topmost branches of the canopy of the oak as I cross the bridge by the lake. The snow has gone. The sunlight drowns the green of the conifers. There is birdsong and signs of life. Snowdrops carpet the earth in patches, white rugs tossed over the brown clay. A song thrush sings from the highest treetop. Then it stops. A moment of transcendence falls. For a time, with the still that follows comes a strange sense of anticipation in the air that drifts and then vanishes, and a true peace reigns.

These are the moments we seek. I sit on the bench and wonder if the awareness of that transcendence comes from sensing the essential existence of the oak, alive and sentient as a clear, pure, breathing spirit unencumbered by others. Born in ancient soils, the oak has dug down into the earth through the centuries and now is deeply embedded here on this patch of earth. I smile and rise. I am starting to think about the essence of being an oak. I wonder if you can become a Green Man in this day and age. Can you step into the oak and become someone other than yourself, someone fused with oak knowledge, with peace and calm?

I wander back into the conifers to seek communion with the goldcrests. Their cymbal chimes still ring in my memory.

And they are there. Their perfect angel calls surround me. From high in the spruce trees, their tinkling chatter falls clear and distinct. Below, beside me, a wren taps out an impatient message into the gorse. Above, two goldcrest tear about amidst the umbellifer heads of the spruce trees. I strain to peer heavenwards into the sun. They are so small. They dart about the bare branches and I see only shades flickering in the violent sunlight. And then they come together, pouring their cries into each other, forty feet above, twisting and turning and falling like thistledown all of a flutter before blowing apart and vanishing.

To hear goldcrests is to be transported to some higher place. I feel physically lighter. I gaze into the spaces over the treetops and feel glee rise within me as an ecstatic fervour that surges through my being.

26 February
Thirty minutes until sunset. Grey skies. A gentle easterly brushes the bushy tops of the spruce trees. A brace of woodcock blast from their bed of bracken, unsettling me with a vivid jolt. I calm and listen and hear, unseen, the thin calls of goldcrests in the trees. They are like the whisperings of wood sprites.

I step off the path into the bracken after the snatched sight of a white bird that cries and flies high above the conifer trees beyond the stream. It must be a buzzard. Yet it was so pale. Like the bird that ghosts over Seven Acre Wood near my home.

19 March

In the firm embrace of the Honywood Oak, I sit and sip tea. The sun, which appeared briefly half an hour ago and poured life into this dull earth, has now been shrouded in grey cloud. The sight of sunlight brought the geese to honk and holler with obvious hilarity into a song of sorts while beneath the raucous rattle there were the fainter notes of robins that fledged about the bracken. I feel the peace I seek. I feel it seep through me, stilling my mind from the noise of thought, emptying my body of anxiety.

The oak is silent. A grey squirrel ploughs headfirst down one of the remaining conifers in the belt to the west. A week or so ago the face of a Green Man peered down at me from the top of the nave in the church at Nayland. In the ridged skin of the oak there are many such faces. And now the dusk chorus rises. Blackbirds and pheasants bring on the night.

22 March

By the time I reach the Honywood Oak, the sun – which has filled the sky all morning – is gone. A crow flies south from one of the upper boughs. I step over the low railing of the enclosure and walk gently to that familiar lee in the west of the trunk, but the cold-feathered fingers of the north wind still find me out as I tuck myself down there. Spring is not yet here but it is not so far away.

My hand upon the oak.

How do you explain that feeling, that rising emotion, which comes as your hand rests against an ancient oak? What is it we feel? Perhaps we do not always appreciate the significance of the value of that touch from one species to another species, from one individual to another individual. How do you explain that sensation which each human being knows when we as one living creature touch upon the life-force of another being of another species?

We know it most keenly as children. We know it best in the stroke of a hand upon a beloved cat or the pat of a favoured dog. Yet we also know it when we see the robin singing daily on the bird table beside the kitchen window. And it is there, too, in the touch of my hand upon the oak tree.

Someone had recently led me to D.H. Lawrence's poem 'Under the Oak' (1916):

For I tell you
Beneath this powerful tree, my whole soul's fluid
Oozes away from me as a sacrifice steam
At the knife of a Druid [. . .]

Above me springs the blood-born mistletoe
In the shady smoke.
But who are you, twittering to and fro
Beneath the oak?[15]

As I read the poem again last night, I thought of Robert Graves's comment that the wren is 'the bird of the Druids'.[16] Is that what Lawrence pictured 'twittering to and fro / Beneath the oak'? Or did he see something of more human form, some spirit of the oak, some glimpse of a Green Man, Woman or Child?

I'd been looking, too, at a Green Man who appears in Shakespeare's *As You Like It*:

Under an oak, whose boughs were mossed with age,
And high top bald with dry antiquity,
A wretched, ragged man, o'ergrown with hair,
Lay sleeping on his back.[17]

The figure beneath the oak is discovered by none other than someone called Orlando. Virgina Woolf's Orlando would have been more likely to have been under the oak himself. Shakespeare's Green Man turns out to be Orlando's brother, Oliver. It is yet another transformation in the oak woods.

I sit back upon the bench and the sun warms my black trousers until they are burning hot. It is the first such heat of the year. The sunshine has brought out an early batch of beetles – each is black-cased and no more than two or three millimetres across. Perhaps it is they that the goldcrest are feeding upon in the tops of the pines above my head.

Something stirs. The seasons shift. Winter is becoming distant, banished. Warmth and sunshine now fill the day. And as I write these words a solitary black beetle wanders onto the blank desert of the page. Another lands, folds diaphanous wings and also ventures over the pale reaches of the notebook.

And so I rise and head off north again along the side of the lake to the pine woods, seeking the soft sibilance of the goldcrest.

25 March

Lady Day. Dad's birthday.

Snow-covered lands.

I arrive an hour or so before sunset. All is still. The spring snow has culled all sound once more, has whitened the land, earth just starting to turn green. Winter has returned. The oak sits frozen.

By the oak, the wind blows from the north-east. A jay flies west as I approach, up towards the top of a spruce that draws my eye to the copse of conifers where a pale angel floats among the trees, hovering above the bracken:

a barn owl out early. As ever, the sight is instantly entrancing; a vision that halts the air in the throat and stops time passing.

Jays sow most of the oaks. Somewhere, I have read of a study that showed each jay may bury up to 5,000 acorns a year. In French, the bird is known as *geai des chênes*, the 'Jay of the oaks'. Their forgetfulness ensures their multitudes of buried acorns are not eaten but turn to oaklets.

Thomas Hardy in 'Domicilium' (1916) writes how:

An oak uprises, springing from a seed
Dropped by some bird a hundred years ago.[18]

Hardy's bird was no doubt a jay.

I sit in the lee of the oak tree. A woodpecker calls high over the conifer copse. I watch it wing its way west. Fifteen minutes to sunset. An opaque-eggshell patch of blue sky hangs in the south, an opening in the grey cloud. Too cold to sit, I rise and walk. I step into the trees a moment. All runs before man. On the emerald path, I think I see a hare. Yet the light thickens and it is hard to tell the real from the unreal. I return to the oak and sit back to back with the tree once again in that southeast lee. To the north comes the cooing call of a buzzard out late, circling still, floating above the tree line, framed between the straight limbs of conifers. I see him. I hear him again.

The sun sets.

All day there has been a tugging remembrance of my

father on his birthday. He would have been seventy-four. Leaning here against this ancient oak, I think of him. There is a sacred sense to this space beside the Honywood Oak today. I can remember my father more clearly. I can recall him more richly for being in the oak's presence. For a time, he is here again and I breathe in his memory and am gladdened.

Grey tundra skies. Two orange-billed greylag geese flee east. It is that time of day when birds turn silent in their nests. It is time to seek mine. Time to head home and build a fire. Time to sit and simply be beside the flames. I lean forward into the darkening gloam.

2 April

Sunshine. North-easterly winds freezing.

As I step through the gate, a woodpecker weaves through the ash trees south of me and halts, red cape bright in the spring sunshine. On the path to the bridge over the lake lies an offering: a single white goose egg. I look around but am unable to tell whose egg this is. A greylag squawks.

At the oak there has been a killing. A splay of white feathers shiver in the wind a few feet from the oak. A fox. Last night or the night before. And now a pigeon is no more. I circle the prey with forensic steps then turn to the oak and examine a section where the bark has come free. A pale yellow glue-ooze plugs a series of tiny, needle-sized holes. The north-easterlies blow a clear cold through the lines of limes, stirring the balls of mistletoe high in the trees. The cold numbs my fingers. I nestle back into the niche in the south-west, shrugging off the chill winds and facing the sun as a sparrowhawk slips through the sky heading north over the pines, then cuts east – a sliver, a slice that drops over the lake and is gone.

A woodpecker drills from somewhere under the sun. I see a shade of something beneath the light, a dark silhouette, but can make out no more. From the north, a jay call echoes, throaty as a deer's bark.

It feels so good to simply be as the oak is. Static and yet apparently at the centre of such a throng of nature, which seems to swirl in centripetal motion inwards on me and the ancient tree. I can hear the call of long-tailed tits though I cannot see them.

I move to sit on the bench, freshly placed, north now of the oak with a view back towards the lake, the yew, the estate houses. A crow has spotted the goose egg. When I turn and look again the crow has the egg in its beak and is plodding slowly through the meadow grass to the sanctity of the yew. The clocks have changed. This added daytime feels such a blessing, flooded as it is with soft sunlight.

Moments pass.

Later in the day, I meet up with Jonathan, whom I have not seen for a while and who drives me over to the oak he calls the Constable Oak, out on the avenue that leads south and away from the coach house. He wants to show me something in the bark.

'They're normally around head height,' he says.

And sure enough, there they are. He points to a D-shaped hole in the bark, six feet or so from the ground, then to another.

'The final nails,' he says.

And they do indeed look like the holes that would be made by old-fashioned nails being hammered into the oak. Each is the exit hole of an oak jewel beetle, which burrows its way through the bark into the sap-wood and brings disease into the already declining oak tree. Jonathan stares up at the shrunken stag-head that tops the tree. Though he does not say anything more, his concern for the tree is clear. There are no obvious signs of infection, no seeping black wounds that signal disease and stem-bleeding. We peer in and see that the edges of these holes are worn, rather than sharp and fresh, which reassures Jonathan a little. They are old holes.

We bid farewell to the Constable Oak and return back across the lanes of the estate, crossing the bridge over the lake towards the Honywood Oak. It is framed glo-riously in the summer sunshine. I jump down from the truck. Jonathan has something he wants to collect from the coach house and returns a few moments later with a white sheet, which he lays on the ground beneath a lower bough of the oak with a perfect sense of ceremony, as though he were some Druidic priest. And then with further echoes of some ancient pagan ritual, he proceeds to shake and beat the oak branch while I stand beside him and watch the collection of insects and bugs that fall and gather into the pale embrace of the sheet.

He halts. We kneel and lean in to see the treasures before us. One is a round-bodied spider whose pea-shaped bulk is painted precisely the same emerald shade of green as the young oak leaf. It is just one of the many thousands of species of invertebrates in Britain that rely upon oaks. More than any other tree species, the oak provides food, shelter and a variety of habitats for vast populations of spiders, mites, centipedes, beetles, wood-lice, ants and any number of other spineless animals. Of all these invertebrates, over five hundred are dependent on the oak in some way. Each ancient oak tree offers a stable and secure environment across hundreds of years for populations of these glorious beasties to live in. Some prefer the leaf litter, others the crevices of the bark, while many more fauna live their lives in the decaying world of the red rot, the rotting heartwood that builds as the oak ages.[19] Before us are a small smattering, a sample of this wealth of insect life which the tree nurtures. We gaze upon them. There are oak weevils and click beetles and chafers, but there are no oak jewel beetles, which is a blessing and a relief. Each creature wanders the vast white expanse of the sheet, unveiled against the blank-ness, suddenly dropped into a desert landscape to roam awhile, utterly naked and bewildered. We carefully lift each one in turn and stare down with Gulliver eyes as they crawl on our skin. Then we return each individual to the home comfort of the oak.

The meeting with the Constable Oak reminds me how many painters and artists have appreciated the oak, and none more so than Jacob George Strutt. In his glorious, paving-slab-sized *Sylva Britannica* (1822), Strutt presented a series of 'Portraits of Forest Trees distinguished by their Antiquity, Magnitude or Beauty'. Each has been 'Drawn from Nature' and then etched by the author. Most of the drawings are of oaks. In his accompanying commentary, Strutt's delight in oaks is evident:

> The oak, admirable alike for its beauty and utility, has ever been distinguished as the glory of the forest; over all the trees of which it may be considered to reign with undisputed sway, both in importance and longevity.

Strutt delves into historical reference, declaring how 'the oak was held sacred by the Greeks, the Romans, the Gauls, and the Britons' and that 'its consecrated shade was devoted to the most solemn ceremonies of the Druids'. Clearly drawing a direct line from the Druidic practices of the past to his own reverence for the ancient oaks, Strutt proudly states that 'scarcely is it held in less veneration by their descendants'.[20]

One of his subjects is the Great Salcey Forest Oak of Northamptonshire, which measures some forty-seven feet in circumference and so, by the calculations of a Mr H. Rooke, Esq. F.S.A., 'cannot be less than fifteen hundred years old'. Strutt's etching shows a vast, hollow shell of an oak beside which impressively antlered deer sit, sheltered by the flourishing and verdant canopy of the very much still alive tree.[21]

In the first volume of his *Remarks on Forest Scenery & Other Woodland Views* (1791), William Gilpin praised other majestic qualities of the oak:

> We seldom see the oak, like other trees, take a twisted form from the winds. It generally preserves its balance; which is one of the grand picturesque beauties of every tree.[22]

Whether bold and upright, or old and knurled, the oak tree was always held in high esteem.

In Britain, the oak tree has long served as a symbol of national identity. The expression 'hearts of oak' exemplifies the essential strength of the oak, a natural state Britain has seen as evocative of its own citizens. An oak tree has even saved an English king.

In 1651, following his defeat in the Battle of Worcester, Charles II was on the run. By September that year, he had taken refuge at Boscobel House in Shropshire with the loyal Royalist Colonel Careless, who declared the house unsafe, so the two hid during the day in an old oak nearby. In his *Arbores Mirabiles: or a description of the most remarkable trees, plants and shrubs in all parts of the world* (1812), Joseph Taylor took up the tale:

> It happened that whilst the king and the colonel were in the tree, a party of the enemy's horse, sent to search the house, came whistling and talking along the road. When they were just under the oak, an owl flew out of a neighbouring tree, and hovered along the ground, as if her wings were broken, and the soldiers merrily pursued it, without making any circumspection.

Taylor also gave a translation of the Latin inscription engraved in the brass plate to memorialise what became known as The Royal Oak:

> Sacred to Jupiter is the Oak.
> This most glorious tree, which,
> For the asylum of the
> Most powerful King Charles II
> God the Greatest and the Best,
> Here caused to flourish.[23]

The Royal Oak rapidly became a tourist site whose visitors stripped it of bark and branches to keep as souvenirs and, soon after, of life.

In the naval history of Britain, 'hearts of oak' was employed as a term not only for the ships that secured the empire but the Britons who sailed upon them. David Garrick's opera of 1759, *Heart of Oak*, spelled the matter out in its chorus:

Heart of oak are our ships,
Heart of oak are our men;
We are always ready;
Steady boys, steady;
We'll fight and we'll conquer again and again.

And the ships of Britain were indeed made of oak. They always had been. The hollowed-oak canoes of the Stone Age had evolved somewhat, but their later embodiments were still being made of oak up until a century or so ago. Horatio Nelson's HMS *Victory*, which led the naval defeat of Napoleon at Trafalgar in 1805, was constructed from some 6,000 trees of which some 90 per cent were oak.[24]

When each boat or ship was taken out of service, the boughs of oak that had been carved and shaped to form the vessel were reused to form the bones of the skeletons of homes all around the country. My own cottage, built for farm labourers sometime in the sixteenth century, is made up of lengths of oak that in their strange holes and joints tell of their life before as parts of a previous incarnation sailing upon the seas and oceans of the world.

That history of oak in shipbuilding reminds us that

only in the last century or so has oak become less of a practical necessity. In a similar way, it is only in the last few decades that oak and wood more generally have become less than a vital fuel for the hearth of every home. Only so very recently have we stepped away from oak trees being essential elements to our daily existence.

8 April
Early afternoon.
A cold wind still blows from the north-east so I tuck down beside the tree in the south-west and peer into the fissures of a piece of oak bark that has fallen from on high. I turn back to the living trunk of the oak and start to steal around it like a treecreeper, circumambulating sunwise around the oak, feeling the skin of the tree on my own.

The variance in the bark of the oak becomes tangible – in places it is pale and stretched until almost beech-like over the scars of previous branches lost to time and season. My fingers brush across the rough ribs of the oak's outer coat. To the east, a section of inner tree is exposed. Dry red heart rot falls from the scar-hole of an old limb lost long ago. I gather some of the dried flesh and it is light in the hand and layered with a complexity only to be beheld when seen up close. I peer in, seeking beetles and bugs.

On the outer reaches of the oak branches, the buds are starting to form brown heads scaly as the skin of lizards. Signs of spring. I creep round to the east where there lies an old, leathery fungus and think how strange that I had never really noticed it before. I reach beyond, to where a black sore weeps a tar-like substance from a bacterial infection of the oak. And above it, the smiling face of a Green Man appears in the whorls of the bark and peers down at mine.

16 April
Spring at last.
Flares of daffodil across the roadside.
Great splashes of primrose.

Magritte clouds drift over from the south, buffeted by a fresh breeze. The remembrance of last sitting in the lee of the oak is as from an age ago, though less than two weeks have passed. The seasons have changed guard. Still, warm spring air has replaced the chill winter winds that blew from the north-east.

From the top of the oak, a chaffinch beckons all with his righteous call. An answer echoes back from some-where over by the lake and he is gone. A fluttered song rains down from three goldfinches that are so soon flee-ing south. And the chaffinch is back, though silent now, and catching flies every once in a while in the spaces between the fingers of those faraway branches.

23 April

Spring sunshine brings seasonal sounds.

There is the buzz of a spring bumblebee woken from a long, deep winter sleep – a gyne or young queen who is soon to become one of this year's queen bees. She is freshly roused by the sun and tears past on an air path seeking a new home.

A chiffchaff calls its name. I approach the Honywood Oak past pairs of preening greylag geese. It is the sunshine of spring that fills the air. Time passes with such gentle haste when sat beneath these oak boughs. A pheasant plods with infinite care about the edges of the oak. A dunnock hops about the mulch chippings of the earth floor then flees for the bramble, vanishing into the background shades. A buzzard rises unhurried in widening circles over the trees to the north.

The ground beneath the oak is soft. I lie and close my eyes. Spring murmurs about me. I feel like I may sleep while all awakes.

26 April

Three days on, icy winds return from the north bringing cold rain in flurries. From under the cover of the evergreen branches of the yew by the lake I sit and watch the April showers. The geese have wandered away from the lake today. Two greylags potter and probe about the base of the oak, divining delicacies from the bark-littered floor. Gusting winds buffet the tops of the conifers, the hedge-like growth at the top of the oak. The winds run across the lake tearing ripples from the surface of the water.

An hour on, I touch the first sprays of green leaf that are now there at the tips of the oak branches where a week ago there were brown buds. Above there is the sibilant mewing of a blue tit that I watch while leaning on the frame of the wooden boundary rail. Then it is gone. And a minute on, the same creature appears again from the dark circle of a round black hole on the east side of the oak. Too small for squirrels to delve into, it might just be a perfect nest site.

Within weeks the tree will be crawling with caterpillars – the favoured food of blue tit chicks. I meet Jonathan Jukes at the oak and he explains to me the precise timing which the birds practise so that the hatching of the blue tit young fits precisely with the appearance of oak-based caterpillars like the green oak tortrix moth caterpillars (*Tortrix viridana*), which emerge each spring and feed on the fresh green shoots of the tree.

'It's a delicate balance. If timings go awry one year, the population of blue tits can collapse,' he says. 'And without the blue tits to eat the caterpillars, the oaks will lose much of their fresh leaf growth. In the worst cases, oak trees can be stripped naked by sudden explosions of tortrix populations.'

The ecology of each oak tree is finely poised. Of those thousands of various invertebrate forms that turn to oaks for food, these tortrix caterpillars are one of many that may upset the equilibrium. Some years, invasions of caterpillars will have dramatic consequences both for the oak trees themselves and for the rest of the populations of creatures which are reliant upon them. Back in the spring of 1785, Gilbert White – vicar of the village of Selborne and progenitor of British nature writing – was complaining of just such an incident.

'Most of our oaks are naked of leaves . . . having been ravaged by the caterpillars of a small *phalaena*.'[25]

30 April
Spring sunshine. Patchy white clouds.

Green leaves are unfurling. The fresh shoots of the horse chestnut trees that line the avenue through the back of the estate fold out like gloved fingers.

On the oak those first flickers of green have grown between strings of spider silk. Blue tits flit about. I watch two chasing each other in tight circles around the vast, severed north limb. They disappear into the darkness of a hole in the bark. In the mulch below them lies a white eggshell that is not theirs.

I meet up with Chris Gibson and his wife, Jude, in their flat, perfectly nestled in the roof space of a redeveloped wharf overlooking the Colne Estuary at Wivenhoe. Four floors above the waterside, there are views of sunrise through one window and of sunset through another. Chris is a naturalist who has only recently retired. He worked for Natural England and English Nature for most of his working life. Jude is something of a whizz on the more minute bugs. They have invited me over to talk oaks over onion soup.

'There's a theory that our connection to oaks is embedded from prehistory,' I find myself saying, 'because acorns provided a staple through times of famine. So when we moved away from Mesolithic living – hunting and gathering – and started to plant the first crops, acorns were the food source that could always be relied on even when everything else failed.'

'They certainly can be used,' agrees Chris. 'They contain starch, so you can make bread from them. Full of tannin though. But soaked and ground up, they will make flour.'

'But a pain to do,' I venture.

'You can feed them to animals, of course,' adds Jude.

'It's always struck me,' says Chris, 'that humans have had a spiritual connection to oaks that is hard to explain.'

I know what he means.

The soup is ready but we continue to talk oaks. We talk Pliny the Elder and the depiction of Druids in Britain cutting mistletoe from oak trees with a golden sickle. We talk of how at Dodona and other sacred sites of Ancient Greece, oaks were grown and tended in the heart of the temples. I tell Chris and Jude how the previous week I met up with a Syrian friend of a friend who is an Alawi Muslim – a sect of Shi'a Islam – and how she explained to me over a coffee that in their belief system the oak tree is also fundamental. An oak could be found in the centre of each Alawi graveyard and in the middle of outside shrine areas. I had wondered then if there was some kind of cultural cross-over from these sites on the Mediterranean coast of Syria to the oak practices in Ancient Greece.

'I mean, Greece is just down the road,' I say.

Chris nods enthusiastically. It is certainly plausible. He sensibly suggests the oak might also have acted as a symbol of longevity much as the yew does on many Britain religious sites, often having been planted on pagan sites hundreds of years before Christianity arrived.

'Could well be,' I agree.

'Or one of these cultural cross-overs,' states Chris. 'Or a bit of both.'

Jude has kindly dug out a couple of books that I should look at. One is a vast compendium on symbolic and ritual plants in Europe. The book weighs a ton. I flick through the pages. It is rather intriguing as it has been published by the University of Ghent in Belgium and so offers a continental perspective on such matters.

'I'm looking to find out more about other cultural links with oaks,' I say. 'Like meeting up with that Syrian friend last week. Because we think that oaks are especially significant to Britain . . .'

'Exactly. We think it's our tree,' adds Jude.

'But if you go to Germany, they have similar notions of the oak and national identity.'

I think of Thor's Oak, for example, also known as Donar's Oak, that grew in Hesse, Germany, which the Anglo-Saxon missionary Saint Boniface is said to have cut down in the eighth century. Soon, we are back to the veneration of the oak right across the northern hemisphere, in fact, everywhere that the oak grows across the globe. I remember from somewhere the fact that there are over six hundred species of oak around the world.

In Britain, there are two native species – the English (or pedunculate) oak (*Quercus robur*) and the sessile oak (*Quercus petraea*). The Honywood Oak, like most in Britain, is an English oak. It was the strength in the timber of this species of oak that was vital to British shipbuilding and so essential to the development of the British Empire. For centuries, the English oak was also known as the 'naval oak'.

'Hearts of oak,' says Jude.

Chris and I both agree.

We finish the onion soup and settle into the comfort of sofas. When we venture back to the oaks it is to the Honywood Oak again. I tell the tale of the other oaks of Marks Hall.

'It was one of three hundred ancient oaks that lived there until the 1950s,' I state.

But Chris knows the story better than I do.

'Until she felled them all,' he says.

I am not quite sure I know what Chris means but continue anyway.

'They were taken out for cash.'

Chris has a different take on the tale.

'And hubris,' he states. 'Or, at least, as I understand it.'

I am all ears.

'When Thomas Phillips Price died,' continues Chris, 'his wife was extremely annoyed at his philandering when he'd been alive. Well, he loved his oaks and so when he died, she took it out on the oaks and said, "Chop them down. I'll have the money."'

It is the first I've heard of such a story. Chris can't remember where or when he heard the tale. The story of Mrs Price's revenge might have come from any number of people's mouths.

'It might be worth following up,' Chris says.

So I do. Sometime later, I speak to Jonathan on the matter. He and Chris are old friends.

'It is hearsay,' he says. 'There's no evidence for the story. The common local legend, if you like, is that she thought she was going to get the freehold on the estate when Phillips Price died and when she didn't, she acted out some kind of revenge, which included felling the trees and also having a former Mrs Price exhumed from the churchyard and reburied in Colchester.'

'Right,' I say.

'But there's no evidence she was acting maliciously,' Jonathan concludes.

It's another telling of the same story Chris has heard – gossip on the felling of the Honywood Oaks.

Chris had actually been working at the Marks Hall Estate from the mid 1980s.

'I was working for Natural England, persuading the Forestry Commission to take out some of the conifers that had been planted in place of the oaks.'

He begins to explain how the woods at Marks Hall had a remarkable butterfly population up until the 1960s and how he'd been involved in helping return wood white (*Leptidea sinapis*) and silver-washed fritillary (*Argynnis paphia*) species to the estate.

'And if Mrs Price had left all her three hundred oaks . . .' I start.

Chris sighs audibly.

'. . . Would the butterflies have survived anyway?'

'It is possible,' he explains.

But the loss of the butterflies in the 1950s was part of a wider decline in woodland species as land management fell away across the English countryside from the end of the First World War. The countryside managers had literally been killed off by the war.

'Lack of land management meant lack of coppicing which meant lack of food plants which meant butterflies died out.'

Were the three hundred ancient oaks of Marks Hall still around then they would have been too old to need much management, apart from removing an overweighing limb and perhaps somehow protecting the surrounding grounds from becoming trampled down.

'Another thing about oak trees is that they support more species of life than any other tree,' says Jude. 'Is that true?'

'It *is* true, actually,' says Chris. 'But it's not as simple as all that. For some forms of wildlife, oak is not the best. So take, for example, lichens. Lichens may grow on oaks and certainly may be abundant but the levels of tannin means that oaks have acidic bark, which means that in terms of number of lichen there are nothing like as many as in other tree species. Sycamore has a much less acidic bark and has a lot more lichens associated with it.'

'And things like galls are incredible,' says Jude. 'Apparently, there are over fifty different galls that grow on oaks.'

She is talking about the various small round growths found on the stems, roots and leaves of oak trees that are a result of the tree responding to certain insects, chiefly small gall wasps, puncturing the skin of the oak and laying their eggs. As the flow of sap is disrupted at the wound, a protective shell is formed around the egg as a layer of tannins produced by the oak cause inflammation. The swelling gradually builds to become a gall. The most familiar form appears as brown marble-sized balls attached to the twigs of oaks, each ball containing one gall wasp egg. Then there are other galls – oak apples, knopper galls, artichoke galls and cherry galls, to name just a few. They are another example of how delightfully complex and intricate the ecology of each oak tree truly is.

'And, of course, galls were the main source of ink until not that long ago,' I add.

I'd read somewhere ages ago about how marble galls were crushed and mixed with water to produce a strong cloth dye and writing ink. There were still a handful of them in a cup on a sideboard in my kitchen awaiting experimentation. Later, when I turned back to the matter, I discovered that the *Andricus kollari* gall wasp which produces the marble gall had been deliberately introduced into Britain from the Middle East in the 1830s specifically in order to provide a supply of marble galls to be used to make dyes for the clothing industry. The first marble galls in Britain were found in 1834 on oaks around the cloth-making towns of Devon such as Exeter, Exmouth and Tiverton.[26]

We shift to discussing the specific environmental conditions which the older oak trees offer. I repeat the line of Oliver Rackham – much-missed ecologist and source of so much wood knowledge – that you can't see the matter as simply that two two-hundred-year-old oak trees equal one four-hundred-year-old oak tree. It is more subtle than that. The older tree will have species living within its ecosystem that can't exist in younger trees. Certain species need the heart rot habitat that forms in older oaks as their insides gradually crumble away. It is one of the aspects of the oak that I really want to ask Chris about – the nature of that heart rot and the minute creatures which inhabit that environment.

'So you need the centre of the oak to rot away to create the specific environment for certain creatures.'

'Yes, absolutely,' agrees Chris. 'And the tree needs to rot down in order to provide greater stability. A heart-rotted tree is more stable than a solid one. That's why you make scaffolding out of hollow steel rather than solid steel.'

Chris explains the exact significance of the oak's heart rot.

'If you've got a cylinder of rot that is going to stay in the same place for six hundred years or so, that's six hundred years for creatures to find and settle in that place.'

It is the fact of being geographically static over such a long period of time that enables the complexity of the habitat. Then there is the aspect of the living part of the oak, the still-growing wooden circle that surrounds the decaying centre, acting as a protection for the ever-rotting habitat. That living part of the oak continues steadily to grow out but also provides a remarkably secure world within its embrace. Other trees, such as beeches, produce heart rot, but it is the longevity of oaks that allows for the development of such complex ecosystems and communities of creatures within the body of the tree for hundreds of years. Other species of trees, like beeches, soon become unstable with rot and fall down far sooner.

I ask Chris about the nature of some of those creatures that live in the heart rot of the oaks.

'I mean loads of them don't even have a common name,' I say.

'Lots of them are beetles,' he says. 'Boring beetles. Because they're boring . . . and boring.'

All three of us laugh.

Jude has already told me how she has a special fascination for the more minute beings of the woods.

'Because of my short-sightedness, my myopia,' she explains, 'I can focus on something so tiny more easily.'

'Jude's brought them to my life,' says Chris.

'I do like an oak weevil,' Jude admits.

We drink coffee and talk stiletto flies and minute spiders. Soon, we've moved back to the oaks and to trees more generally and to how in the last twenty years scientists have started to realise more and more about the understorey of the woods – how mycelial networks formed in the soils by delicate fungal growths on the finest root stems of individual trees are passing materials, and even information, to each other.

'You can put a radioactive tracer on a tree and you'll find that radioactivity ends up in a tree further down the way. It's gone down the roots, through the mycorrhizal fungal links and ended up in another tree,' says Chris. 'So the trees are connected. They can communicate chemically. There's a flow of nutrients but also of communication signals.'

He offers an example.

'So one oak can actually say to the others around, "I'm being eaten by gall wasps, get your tannins up!"'

'And everyone in the scientific community seems to have accepted this – that such forms of communication are taking place between trees,' I add.

But if trees thrive best in forests and woods where others can share resources and warn of enemies, then what of the lone, solitary, ancient oak?

'Rather argues against the idealised vision of the single grand old oak on its own,' I say.

'It does, doesn't it?' agrees Jude. 'Like a Capability Brown landscape. It is done because it looks good, not because it is natural.'

'It's a lone oak, a Billy-no-mates oak,' jokes Chris. 'It's got no one to talk to, no one to defend it against insect invasions.'

'It's sad,' says Jude. She gives a little laugh. 'You can start to get a bit anthropomorphic, can't you?'

You can indeed.

I inevitably think of the Honywood Oak. Surely it isn't mere anthropomorphising to see that tree as having been left isolated and alone by the action some sixty years ago of felling the three hundred other ancient oaks. From what we now knew of the science of communication and connections between trees, those other oaks, were they still here, would live a communal existence within the woodland. Instead, the Honywood Oak is all on its own.

Jude starts to tell of her time volunteering with the British Trust for Conservation.

'Building pathways through woods and such like,' she says. 'And one day we were on a training course and we all went on a tree-hugging activity. You had to select a tree – and I'm sure I selected an oak – and you had to go and hug it and stand near it and look at it, look at the shapes and look at the bark patterns and get to know that particular tree. You were there for ten, fifteen minutes.' Jude pauses. 'It sounds a bit wacky and mad, but . . .'

I smile. I know exactly what she means.

It is so hard to explain. There really is something in that 'but' which I understand, something about a connection with an oak tree, something about how in those moments of meditation, of musing on the nature of that single, particular tree, you really do create some sort of feeling for the tree and in the process experience some form of profound peace.

'I love that you've brought that up,' I say. 'Basically what happened with me was that after I'd been sitting next to this old oak tree for a number of months, I did kind of get to know this tree. Then I started to read various wise words of others both philosophical and spiritual on recognising the individual oak, the individual tree – Martin Buber and Gary Snyder, for example – I mean . . .' I am starting to rant a little. I know it and yet I also know that in Jude and Chris, I have fellow believers.

'No one would think you were weird if you were talking about a relationship with a cat or a dog. And I think you can do that with an oak tree. I genuinely think you can.' I laugh to cover what I am saying. 'Then that links back into the veneration of oaks throughout history and how you get spirits and human forms like the Green Man associated with the oak.'

I pause.

'I'm so glad you said that, Jude,' I say again. 'But it's funny how you covered it earlier by saying it was wacky and mad. I've done exactly the same saying it was a bit odd and out there. A bit too weird or woo woo . . .'

We all chuckle.

And it is true. It is hard to explain that sense of connection to an oak tree without sounding suddenly rather more extremely New Agey than I feel I am. And yet everyday folk like Chris and Jude understand exactly what I am on about. There just doesn't exist an appropriate language to describe some of the feelings.

'I mean, an oak is a living thing,' says Jude.

'And even when they're dead, there's something about them. Even with dead oak trees. They have such a presence in the landscape,' adds Chris.

And so we start to wonder whether you would get the same feeling by hugging a dead oak as with a live one. And at that point, Jude sensibly suggests we have another coffee. As the kettle boils, I explain that I want to know more about the Green Man figure, want to examine more closely the sense of peace I have experienced being beside the Honywood Oak. We start to chat about some of the spiritual aspects of oaks. I know Chris and Jude well enough to step into this ground.

'Spiritual might not be the right word,' I hear myself saying. 'It is weird how there is some kind of connection that takes place when you spend time next to an oak.'

'There is,' says Jude. 'It might not be consciously. It's

something more than that. It might be inherent within us . . . It's hard to put into words, isn't it?'

It is. I know what Jude means and I know, too, the difficulty of expressing it. I have started a few conversations recently on the same thoughts and always seem to be adding 'I don't know' as my words fail me.

Jude is thinking again of her experience of the tree-hugging activity on her training course some years back. Its impact is still apparent.

'It really was wonderful,' she says. 'I really can understand it.'

Chris presents me with another book. It is a slim work with the title *Oak*. The book is, in fact, an exhibition catalogue and consists of a series of paintings of a single oak tree.

'Do you know of this?' Chris says.

I do. *Oak* is the printed outcome of the work of painter Stephen Taylor who painted one single oak tree fifty times over a period of three years. I came across his work some years back but have yet to meet him and hear his thoughts on oaks. Clearly, there is much we have in common.

Chris knows Stephen from a time when Chris was working for English Nature and met Stephen at the oak tree and talked him through the nature of the oak from a naturalist's perspective.

'We spent many happy hours standing by it, chatting about it.'

Sometime soon, I, too, will get to chat with Stephen about his oak.

7 May
A warming sun.
There is a lazy, summery feel to the day. It is a week since I was last here by the oak and the growth of new leaf is startling. All is now green. A vivid, fresh, lime-green forest has formed where a week before was bare branch. And as I lean in and stare it seems as though the oak leaves are visibly unfurling. The catkin stems of the flowers have come out, too, drooping down like shrunken, unripe grapevines.

The *tap-tap-tap* of a woodpecker reaches across the warm air from the pines far to the north. In the oak, great tits call. A chiffchaff's call is answered by another across the lake. Tiny black flies are rising from the mulch floor. These are the four-winged gall wasps that will bring oak apples to the branches by midsummer.

Spring is a time of insect invasion for the oak. Some five hundred insect species may live in or under the canopy of the oak tree. Most are parasitic. As soon as those lime-green new leaves start to appear they are pounced upon by an army of micro-moths and wasps and flies. The green oak tortrix moth caterpillars hide and feast in freshly rolled-up leaves held by their own silk threads and can abseil down to the ground on longer ropes of silk at the first sign of disturbance. The oak relies on the blue tit chicks to stuff their ever-hungry mouths with the larvae of tortrix and other moth caterpillars before the caterpillars have eaten their way through the tree's newly sprung leaf growth.

12 May

Patchy cloud in the sky drifts away to the east. A greyer cloud with a cold-fingered breeze pushes in from the west.

The oak leaves have opened and darkened their green hue in the five days since I have been here. The tree unfolds. The oak grows ever larger. Each hour of sunshine brings forth fresh waxy leaf, born to the world with the same crumpled skin as a baby – with each breath the soft, silken skin will open out, will become smooth.

A clutch of young leaves lie at the foot of the oak. There is something remarkably forlorn in their presence, something intensely heart-rending in the knowledge that they will not grow to full size and shape – even though above me a million others stretch and grow greener and will eventually, as autumn creeps in, turn to earthen brown and fall. Still, the sight of those fading young leaves saddens.

I walk to the oak still in wonder at the transformation, brushing the soft leaves with my fingertips before turning to the trunk of the tree and feeling the warmth of the southern sections of the oak's bark against the palm of my hand. I peer in at an array of grey pods that have formed in a fissure in the wood – each pale pod is centred with a dark circle from where a creature has recently sprung. The insect invasion is everywhere. In places the bark of the oak is pin-pricked with the holes of some strange unseen and microscopic creatures. I lean in until my head is so close to the oak I can smell the musky dry dust of green lichen. I peer into fissures where otherwise invisible silk threads hold fragments of spring detritus, the cast-off casings of inconspicuous seed heads.

On the pale exposed wood where the bark has broken away lies a moth almost indistinguishable against the oak wood beneath it, only betrayed in my myopic examination by the splotches of black on the wing. It is a pale variety of oak nycteoline moth (*Nycteola revayana*) whose caterpillars would already have joined the hordes of other larvae feasting on the fresh green leaf of the Honywood Oak.

I have been spending evenings reading about the moths associated with oaks, delighting in the glory of their titles.

Some of the names of moth caterpillars that feed on the oak: gypsy; black arches; scarce dagger; sprawler; clouded drab; twin-spotted Quaker; lobster; lesser satin; vapourer; pale tussock; grey shoulder-knot; dark crimson underwing; light crimson underwing.

Yet there is only one butterfly whose caterpillars feed on the oak – the purple hairstreak.

I sit. After a time, a tiny, light-bodied fly crawls in spiralling circles about my notebook, soon to be followed by another – a darker-bodied creature that lifts in the air awkwardly and passes me at eye-level.

'*Hymenoptera*,' I say out loud with some authority.

It is a gall wasp. Though which one I cannot say for certain, the creature before me is for sure from the order of *Hymenoptera*. That is not difficult to define. There are over 150,000 different known species of such creatures that form the order of *Hymenoptera* including wasps, bees and ants.

I watch the oak from ten yards away, studying the spring activity with the wide-eyed wonder of binocular vision. A bumblebee disappears, Alice-like, into a small hole at the base of the oak.

'*Terrestris*,' I say out loud again to no one and hold my eyes on that dark entrance into the earth.

The bee reappears a moment later, staggering out on weak legs before launching into faltering flight. She is another of those young queens that have recently awoken from their long winter sleep and now fly uncertainly about just above ground level, seeking a place to settle for the summer where they can form their first hive, where they can become queen bee to their colony. *Terrestris* is, as its name suggests, a bee whose homes are often forged from dim, earthen openings.

Later, as the sun falls away, I walk over to the site of an archaeological dig that has recently opened up where the main country house of the Marks Hall Estate once stood. There are half a dozen souls still crouched over the soil. They are like human *terrestris* bumblebees. It is the end of a long day of delving into the earth. My friend Ellie Mead is there with her band of fellow diggers, suitably muddied and enthused at the work they have done, the pieces of the past they have recovered. She talks me through the thirteenth-century mediaeval finds that have emerged from the soil – great shards of pale, glazed Hedingham ware.

'There's something else,' she says.

Her colleague Graham has unearthed an even greater find. Ellie fetches a small box from the wooden shed that serves as dig headquarters and opens it to reveal a lozenge-shaped silver disc inscribed with an as-yet-undeciphered inscription. The pendant dates to a time when the Honywood Oak was a mere sapling.

19 May
An hour to sunset.
Chiffchaff call beneath the bellowing of the geese. Spring sunshine in the early evening backlights the oak, soaking the young leaves in light. The colour of the oak leaves has changed over the weeks – from the duller grey-brushed tints of the first fronds the tree bore in April to the startling emerald allure of the leaf now lit in the low sun.

Midges fidget in the late airs. Gall flies crawl over the oak. A solitary wood ant wanders high above the ground, up and over the ridges and valleys of the dark bark. I peer even closer in. A spider – a mere few millimetres across – traverses an inch-wide web. These are worlds in miniature but universes real and alive and existing in the edges and crannies of the oak tree.

4 June
Mid-afternoon.
The bees stream back and forth. It is the first hot day of the year and the oak is resplendent, stretching out towards the blue sky.

A breeze blows and it is welcome. The pines sway while the oak remains a fixed mark in the earth. Blue tits skip and flit through the air like bats, darting at all angles in fractured lines, returning to feed, frenzied with the feast on offer. A chiffchaff calls. The geese stroll with their young, glaring and hissing at all. An orange-tip butterfly flutters by.

A peacock walks uneasily about and halts in the shade of the yew tree. I have already read Louis MacNeice's poem 'Woods' once today. Now I read again of 'woods packed with birds and ghosts', of the 'neolithic night' and how 'the gate to Legend remains unbarred'.[27] I rise and go into those tame woods behind me.

We all become better beings when we step back into the woods.

18 June
Mid-afternoon.
A storm is due.
Two weeks on and the oak leaf is an even darker green – more so in this dark light of the afternoon with rain so

imminent. John Clare – nineteenth-century rural poet supreme on birds and woods and trees – wrote in his poem 'Summer' how:

> The oak's slow-opening leaf, of deepening hue,
> Bespeaks the power of Summer once again.[28]

I feel that surge with the approach of summer. My head is light. I am joy. Anything is possible in the world. The future calls.

All sits in a sultry purgatory awaiting the storm. When it breaks I shall settle back down into that winter seat in the lee of the oak that held me through those cold days.

Midsummer's Eve is so soon approaching – a matter of days away. The year slips by. I await the rain and wonder on odd conundrums. What is the life of any human to the oak that lives on through the lives of so many generations? We are as bees, or birds, that come and go across the years and pause a second here beside the boughs before we wander on. This great oak has existed across the time that some thirty generations of human beings like me have been born and lived their lives and died back into the soil. And still the oak lives on.

For many years I have kept a postcard of Titian's *Allegory of Prudence* on my desk, which I have had since I was a child, picked up one day on a visit to the National Gallery in London where the painting hangs. The picture consists of three faces of a man at different stages of his life – youth, maturity and old age. Beneath the three faces are a dog, a lion and a wolf. The same triptych of ages is often seen in the oak – the young sapling, the flourishing, green-crowned tree and the stag-headed mature tree.

John Dryden expressed that three-staged ageing of the oak in his poem 'Palamon and Arcite', first published in *Fables, Ancient and Modern* (1700):

> The monarch oak, the patriarch of the trees,
> Shoots rising up, and spreads by slow degrees;
> Three centuries he grows, and three he stays,
> Supreme in state, and in three more decays.[29]

In J.C. Shenstone's *The Oak Tree in Essex* from 1894, I had found reference to the 'Oak's Pedigree' which Shenstone said 'is scarcely an exaggeration' concerning an ancient oak's age:

> In my great grandsire's trunk did Druids dwell;
> My grandsire with the Roman eagle fell;
> Myself a sapling when my father bore
> The hero Edward to the Gallic shore.[30]

Thinking of an oak's ancestry soon takes you back into prehistory.

19 June
Early morning.
A gauze of mist. Rain in the air.
I woke at dawn to a feeling of being drawn back to the oak.

So now I am here. Peacocks cry from the walled gardens. Their calls echo, seeping through these saturated airs to fall in waves over the oak. Midges and mosquitoes rise in this misty early morning air. One falls. It lies with crossed legs still twitching in the breeze, a pile of banded segments of stick.

Chiffchaff and chaffinch fill the higher airs. Below is the chuckle of corvids clustered somewhere beyond, a covey by the burnt oak. The scratchy voice of a white-throat in the pines opens out to fluid song and from somewhere in the far distance comes the call of a willow warbler.

I am learning to settle into silence. I am learning to listen.

Last night, I came across 'The Old Oak Tree' by W.H. Davies.

> I sit beneath your leaves, old oak,
> You mighty one of all the trees;

Within whose hollow trunk a man
Could stable his big horse with ease.

I see your knuckles hard and strong,
But have no fear they'll come to blows;
Your life is long, and mine is short,
But which has known the greater woes?

Thou hast not seen starved women here,
Or man gone mad because ill-fed –
Who stares at stones in city streets,
Mistaking them for hunks of bread.

Thou hast not felt the shivering backs
Of homeless children lying down
And sleeping in the cold, night air –
Like doors and walls, in London town.[31]

Davies was a vagabond poet of the early twentieth century who lost a leg jumping trains in the States. His was a gentle wandering philosophy. Yet the romantic vision of 'The Old Oak Tree' is contrasted with the destitute urban poor – the 'starved women', the 'homeless children' and the 'man gone mad' – struggling to survive. The oak offers Davies a freedom from the ways of the world, a place of shelter and peace.

From some long-distant conversation, I remembered how once my father had said he'd always liked Davies' poem that begins 'What is this life if, full of care'. Dad wasn't really a man of poetry and so the connection with Davies had stuck firm.

Last night, I sat beside the fire and saw Dad's face again and felt his presence.

> What is this life if, full of care,
> We have no time to stand and stare.
>
> No time to sit beneath the boughs
> And stare as long as sheep or cows.[32]

That is what I am learning to do.

27 June
Late afternoon.
House martins soar above the oak. It is close, muggy summer weather and flies are about, filling the air with black specks. A grey squirrel steps warily on the mulch floor and flees as the noisy clatter of human voices drifts over from a wedding at the coach house, breaking the soft orchestra of natural sound with a spray of clashing, shrill notes.

Tucked into the lee of the tree, I sit with my back resting against the oak. The day is hot. I close my eyes. Two chiffchaff play out staccato calls and above all other sound, almost too high to hear, there is the sibilant *wheet* of a goldfinch. Some moments later, I open my eyes to the flicker of a grey micro-moth no more than half an inch across that flies a foot before my face, precisely camouflaged against the pale lichen growing on the bark.

29 June

The brown form of a kestrel flies past at eye-height.

I meet Jonathan and together we step over the wooden rail that surrounds the Honywood Oak.

'The area will shortly be hand-weeded,' he says.

He wants to gather up any oak seedlings beforehand. We search. There are tiny sycamore and holly springing from the mulched earth but no oak seedlings to be seen.

'Here's one,' he says, finally.

And now as we look, we find half a dozen more. They are the next generation, the offspring of the Hony-wood Oak.

We walk away and both look back at the oak and talk of how the oak has flourished since the opening of the shelter belt. Jonathan sees the green of the leaves. The dark hue of the oak leaves is a sign of health. I think of the trees on Two Oak Hill. One with dark green leaves, one pale.

'The canopy's much more symmetrical now,' I say.

Jonathan agrees. He says he has plans to take out more of the sycamores and the pines immediately to the south to provide even more space and air and light for the oak.

'For seven hundred years or so, that oak didn't have much around it,' he says.

That space and light has gradually been encroached in the last century or so. Now, the Honywood Oak is being given back the room to grow. It is so clearly grateful.

Jonathan heads off. I sit on the bench and simply watch a while. A chiffchaff calls. A green woodpecker flies. The skies are greying. Rain is due in an hour or two.

I step in close to the east side of the oak and run my hand against the pale exposed cambium, dotted with different-sized dark exit holes of beetles. I hold my hand there until the wood warms. Then I step away.

In his brilliant essay *The Tree* (1979), John Fowles writes how we sense we are closest to a 'tree's essence' when we come across that tree growing in glorious isolation. Something of their standing when isolated – like the Honywood Oak – seems to entice us into a belief that we can best gain some knowledge of the tree in such a state. Yet, as Fowles notes, trees are born to live not as lone sentinels but as social beings, in woods and forests.[33] The recent scientific developments into the intricate ways of interaction between trees on microscopic levels of connectivity, through mycelia networks consisting of meshes of fungal root systems, only serve to enhance the argument that trees really thrive in societies alongside others with whom they may be cared for and protected, with whom they may communicate and exchange vital resources. What science may state to us now, many have long known to be true.

In *Far from the Madding Crowd*, Thomas Hardy tells how:

> The instinctive act of human-kind was to stand, and listen, and learn how the trees on the right and the trees on the left wailed or chanted to each other in the regular antiphonies of a cathedral choir.[34]

Trees speak to each other. They thrive in communities of trees.

And while this is certainly true, I wonder on the solitary ancient oaks that pass their days apparently separate from the society of others. Like the wise old souls, the aged pilgrims, philosophers and saints of our species who choose to extract themselves from the ways of society in order to think more deeply of the meaning of existence, we give those individual trees like the Honywood Oak special properties, endow them with particular significance. They are the oaks we most venerate.

Edmund Spenser in *The Shepheardes Calender* (1579) captures exactly that essence:

> There grew an aged tree on the green,
> A goodly Oak sometime had it been . . .
> For it had been an ancient tree,
> Sacred with many a mystery [35]

2 July

Chiffchaff calls cease suddenly as a gentle rain starts to fall, filling the airs beneath the oak with the soft sounds of the shower – a leaf timpani. The leaves are still green, still young, and the tone is light. In autumn, rain drums on the aged, brown leaves with a tinnier echo.

At the base of the tree a ground beetle makes a path over the bark, across the ravines and caverns, the earthy shade of its body slipping into perfect camouflage beside the discarded skins of pine kernels left by a squirrel.

I watch red-tinged wood ants creep across the woodchip. All falls still. I peer into the eyes of a woodlouse then turn to the sound of scratching on the south side of the oak. I do not rise but halt and wait. A leveret lopes into view. Black-tipped ears are held high; dark amber eyes flicker and shine. The young hare steps towards my shrunken shape, folded there against the oak tree and at eight feet away the creature halts and stares and I feel the stare and look back. And the hare turns and vanishes out of sight behind the oak, only to walk widdershins about the tree and appear again; and then heads north, beyond the bird's-foot trefoil and the seated mallard ducks, into the tall meadow grasses.

8 July

Warm summer sunshine.

This morning on my way towards the coach house, on the road into the estate a dark snake crosses the tarmac before me and slides away into the longer grass.

I realise how much darker the oak is now. It is verdant – fecund and fertile. The miniscule ball-bearing acorns have grown – doubling in size in a week. The leaves are deep green. The late afternoon summer sunshine frames their healthy lustre. I watch the shadow play as the silhouette of leaves above flickers across the surface of others below. The leaves are sticky now, each coated with a thin veneer of resin tacky to the touch of my fingertips. It is honeydew. Though the bees nesting in the south of the oak seem to show no interest in harvesting it from the leaves for honey, some bees will and when they do, they will produce oak honeydew honey that is treasured for its sweetness and its health-giving properties, for it is abundantly rich in antioxidants.

Two thousand years ago, Virgil wrote in one of his pastoral pieces or *Eclogues* how 'the sacred oak is a-drone with bees'.[36]

16 July
29°C.
Sweltering now, the summer weighs heavily.

A grey moth vanishes into the bark of the oak in perfect disguise. It might be an oak processionary moth (*Thaumetopoea processionea*) but I cannot tell for it has truly disappeared against the background of the bark.

A weariness is evident in the birdcalls. The muted tones of the chiffchaff are half-hearted; the chaffinch is discordant. At the lakeside, the geese are silent for once. At the Honywood Oak, that same slumberous feel fills the air.

The oak seems quiet, only stirred awake by the welcome breezes that waft the honeydewed leaves. From their entrance hole in the south, a ponderous stream of bees slips through the hot air. One lone creature – disorientated and distraught in the heat – flies up and down the vast trunk seeking a home that only lies some six feet away. Speckled wood butterflies float brown-winged among the dock leaves, rising and falling with silent grace. All is still. Then something stirs the geese to call.

PART III

BEING WITH OAKS

For a year or so, the Honywood Oak became somewhere I went to step away from my life. The quiet observation of the oak and the creatures that lived within its ecosystem offered a practice that kept my mind happily occupied. The world of the oak provided a safe place which was calm and peaceful and where I could flee from thoughts about my fracturing relationship with my partner or anything else at all in my life that should have been the focus of my attention. It is perhaps not so strange that I went there. I earn my crust teaching others to go into the wild, to experience and appreciate the natural world. To be outside in nature has always acted as an elixir to me.

Then I stopped going to the Honywood Oak. It is only recently that I have realised why that was. I fell in love again. Some years later, when the fires of that love started to fade, I found I could do nothing but turn back to the oaks. I grew daily more anxious within my own skin. I now lived alone. I found myself increasingly isolated from others. I found a comfort in being by oaks. Yet the Honywood Oak was a drive away. I turned to other oaks that were closer to my doorstep. There was an oak in the field behind my home. There I would simply sit and stare up at the tree. It was no ancient being like the Honywood Oak but it was an oak all the same and one that I could walk to be beside, one that was a hundred yards from my back door. I began to call it the Field Oak.

And there was another oak. A mile or so from my home was a rise of land I had come to know as Two Oak Hill. One of the two oaks that rose from that patch of earth was a stag-headed oak which was easy to climb, there being a ladder of branches that enabled a simple step up to a convenient seat upon a bough some twelve feet from the ground. Whenever I could, I took the chance to climb from the earth into the Stag-Headed Oak. Things are very different from within an oak tree. Life seems simpler. Everything is more peaceful.

I became ever closer to oaks. I made being beside an oak tree part of my daily ritual, like a religious practice. Rather than merely visiting the Honywood Oak on the way to or from work I now sought to spend more time each day in that still world beside an oak tree. I would turn from all that troubled me in life and work, in order to sit beside an oak and focus my thoughts on the presence of that tree. By an oak tree, my mind became calmer. It was a form of meditation. I spent more and more time in the company of oaks. I leaned on oak trees for support. In many ways, it was a stepping away, a withdrawal from my own species. I can see that now. The silent sentient ways of the oak became something I strove for, as though they were personality traits that I could take on if only I spent enough time beside or within an oak tree. It was a deeply meditative way of being that I searched for. I believed such a state would keep me sane. Whether I was fleeing from being human, or seeking to be of the tree, I still do not know. At times, it felt as if I was drifting too far, like sailing too far from the shore.

And I did turn, at times, to the thoughts of others. I sought out those sympathetic souls who kindly offered me their own wise words on what oaks meant to them, how it was that oaks offered us peace and calm and even what it meant to be the oak.

16 February

Last night, stepping outside into clear darkness, I heard a little owl (*Athene noctua*) call from the south. The cat-like plaintive cry came from beneath the flickering light of Sirius the Dog Star.

This morning, the day breaks across the fields before I have risen from my bedside. Some moments later, I catch sight of the dawn light reaching low over the frost as I peer from the cottage window and watch the breaths of warmed air lifting in rolling clouds from the frozen ground.

In the afternoon I walk from the Field Oak down towards the church to watch the sun set. The late sun

blinds so I turn east to where the moon lies pale and gibbous, high in the sky. It is not yet five. As I step out and
down the road I think how each individual human child
will grow and be quite their very own being. And then I
think how each oak tree also has its own individuality, its
own essence in quite the same way, too. Each oak has a
distinctiveness which may be seen, felt and known – as
with my own children, as with every human that lives
upon this earth. I have been re-reading the existential
philosophy of Martin Buber on trees:

> I contemplate a tree.
> . . . the tree remains my object and has its place and
> its time span, its kind and condition.
> . . . as I contemplate the tree I am drawn into a
> relation, and the tree ceases to be an It. The power
> of exclusiveness has seized me . . .
>
> Does the tree then have consciousness, similar to
> our own? I have no experience of that. But thinking
> that you have brought this off in your case, must
> you again divide the indivisible? What I encounter
> is neither the soul of a tree nor a dryad, but the tree
> itself.[1]

Today, once more, I read Buber on the true observation
of trees and then spend an hour and more contemplating
that oak in the field behind my cottage to which I am
ever drawn these days. I sit and focus on the form of

that oak. I only realise some time later that what I am doing is a form of meditative practice. The focus on the oak allows me to still the frantic thoughts in my head. I sit and transfer my concentration to the oak. I consider all that made the oak an individual. I see the shape and structure of the bare branches of the oak. I see the darkness of a tear in the bark. I see a golden oak leaf that still clings to the tree.

And after time being by that oak tree in the field, I feel so much calmer. I rise and walk away along the country lanes to the south before heading west to watch the light of the day fading to night.

An hour on, and I notice now how the waxing moon has grown more full, more gibbous, since the afternoon. In this deep dusk, silence seems to sit so easily. It will lie more solidly still in the night, but for now I feel the land gently sinking into silence as the dusk birdsong fades away. It seems an age since I have sat in moonlight. The Field Oak is iron black. The wire branches are crazed and chaotic before the open spread of stars filling the firmament. It is not so cold. The wind has lessened. Layers of cloth hold the night's air from my skin. My hands are warm though bare. A certain cold pinch to my nose is all I can feel of this winter's night; a touch, nothing more. It is the quiet of night that I feel most heavily upon me and on the lands about as they settle towards sleep.

It is a clear night and when I shut the torchlight off, I see how the moonlight has grown brighter. The oak trunk is now lit with a leaden, silver sheen that is broken by the black shadows of branches caught in the metallic light. I go in from the field. By the break in the hedge before the green lane, I turn back and see the oak cast in a mercurial blanket, and as I reach the back door of the cottage, I realise it is a lunar shawl, pale as death, that covers the lower body of the oak.

19 February

It can take many years of being in a place before we are finally able to see what there is to see.

Gary Snyder – the poet and environmental thinker – writes in *A Place in Space* (1995) of walking past an old oak for twenty years before one day he 'actually paid attention' to that specific oak and 'felt its oldness, suchness, inwardness, oakness, as if it were my own'. It is a moment of revelation and of intimacy. As Snyder says, there is a dramatic sense of 'suddenly meeting the oak'.[2]

It is much the same way for me and the Field Oak. Only recently have I given the time to pay attention to that specific oak, to contemplate the individual essence of that oak. Now I understand what it is to see each oak for its own oakness. It is not hard to do. It merely takes a little time. Look at the oak tree before you. See the pattern of the trunk, the branches, the canopy of leaves. Then you will start to see the details of what makes this oak tree different from every other oak tree that grows upon the earth, which has ever grown upon the earth. Each oak tree is unique, just like each human being.

The morning is an explosion of glory – the sun rising vast

over the fields from behind the shadows of Seven Acre Wood. Chaffinches scream from the highest branches of the apple trees. I am drawn outside again by the beauty of the day and head for Two Oak Hill. I step onto the summit of the hilltop where the sun casts a slanted light, half-risen into a winter sky made pure once more. Beside the two oaks that stand upon the western edge of the rise, I halt. I touch the skin of one, lean against its body and gaze down at the world below and the unfolding day. These two oaks are no ancient grandees. They have grown from jay-sown acorns for perhaps two hundred years or so, not much more. And yet they hold such a presence here that I return often to rest beside the body of the furthest oak and watch the world a while from on high. The lands here are formed of sandy, fine glacial moraine that erodes downslope with wind and rain and has left each oak's roots exposed. A sacred stream runs away south, skirting the lower reaches of the hill, and has gently done so over so many years that it has formed a shallow valley. A twin hilltop over to the west is topped by the stone tower of a church with equally fine views from a bench that lies against the southern wall. Yet I prefer to sit here beside these oaks.

29 February

I rise at dawn this morning and walk to the lane and stand behind the line of hedge oaks as the sun rises into a clear sky. The ground still holds the pale touch of frost. Above, a song thrush sings a stuttering string of song from one of the higher branches of oak. I stand and listen for a while, watching the bird, spellbound, then wander back the hundred yards or so to the green lane and the field. The moon, whose appearance in the evening sky last week everyone has commented on – its swollen pregnancy, its swirl of colours: pinks and mauves and oranges – now sits, a perfect sphere, pale above the Field Oak. And I do wonder what those ancient people, whose feet walked these lands so many years before, made of that vast circle in the sky. They lived in nature so much more than we do now. They felt the nature of the world about them.

I come and sit for a few seconds – it feels no more – back beside the Field Oak whose upper limbs are held in the new light of day – clean and fresh – until I am forced by time to head inside, to wash and get ready for the day of work. Yet even just to sit for a few snatched seconds, to halt the day and watch and listen in silent reverence, is blessed bliss.

Yesterday, I watched film footage of a harvest scene from long ago, as the last sheaves of corn were lifted onto the wagon. A branch of oak leaves was placed upon the gathered harvest before a maiden climbed aboard, a wreath in her woven hair. You would make the corn dollies from the last scythed cuttings of corn, my mother told me. It is another dying piece of country lore.

6 March

Dawn. The sun is rising through a splash of orange and mauves. From the copse beyond the field a buzzard curves upwards. The oak seems to stretch and spread before the rising sun.

Snow was due to fall this morning. The sky is blue and clear and only frost whitens the ground. The vestiges of snowflakes are frozen on the wooden railing, untouched as yet by sun. In the field, the oak awakes. I walk through meadow grasses stiff and cold. The camping chair which I leave in this cleared space is frozen; a round pool of ice has formed in the seat. So I creep closer to the scrub and bramble and watch the sun creep round, too, steadily lighting more of the land, touching the northernmost branches, the crown of the oak. Sunlight seeps in. Life returns to the field in coos and cheeps and chirps and distant barks, in the natural noises, the voices, of dawn. And I notice how the oak saplings that have sprung from the earth and spread across the meadow, rising through the uncut grasses, have held on to their oak leaves all winter and how those oak leaves shine so bronze.

I step back and over the green lane at the back of the cottage and walk down the road to the field behind the hedge of oaks, the better to watch the sun rise.

And then there is the most extraordinary moment.

I catch a glimpse of the strangest sight. It is a creature of some sort, which flees down the road, towards the east. Yet it is in human form – long-haired, a young woman or a child perhaps and dressed in green, camouflaged for forest leaves and trees though not against the tarmac of the road. For an instant, I am bewitched. I don't know what to make of so strange a figure. I wonder if I've really seen it at all. Yet this dawn creature seems too real. Its flight has even set a dog barking.

I go back into the field and wish I had followed whatever it is and so run down the road after it as I try to get a second glimpse. Only now it is too late. I know it is. The road is empty. There is not a sign of anyone.

The figure has gone and I am left wondering what the hell I have seen. I really do worry that I am hallucinating. I have become so withdrawn these days. My mind is not so solid, not so secure. But it is true. I did see something. I did see a glimpse of someone. Or something. And the dog over the road did start to bark. It is still barking.

Did I really catch sight of a Green Man running down the road? I laugh. No. I see the creature once more in my mind's eye – the slight frame, the long brown hair. No. It was more of a Green Child than a Green Man. And I think immediately of the tale of those two Green Children who were found beside a wolf pit in a village not so very far from here many centuries before.

Later, when the day is gone, I turn back to the thirteenth-century chronicler Ralph of Coggeshall, whose account of the Green Children is eight hundred years old. I know I've mentioned these strange creatures before but something keeps drawing me back to them. The account of their appearance is just so enticing.

Another wonderful thing happened in Suffolk at St. Mary's of the Wolf-pits. A boy and his sister were found by the inhabitants of that place near the mouth of a pit which is there, who had the form of all their limbs like to those of other men, but they differed in the colour of their skin from all the people of our habitable world; for the whole surface of their skin was tinged of a green colour. No one could understand their speech.

The siblings were thought to come from another world. In time, the sister apparently learnt to speak English and so eventually told of how they had ended up in that Suffolk village.

Being frequently asked about the people of her country [the girl] asserted that the inhabitants, and all they had in that country, were of a green colour, and that they saw no sun, but enjoyed a degree of light like what is after sunset. Being asked how she came into this country with the aforesaid boy, she replied, that as they were following their flocks, they came to a certain cavern, on entering which they heard a delightful sound of bells; ravished by whose sweetness, they went for a long time wandering on through the cavern, until they came to its mouth. When they came out of it, they were struck senseless by the excessive light of the sun, and the unusual temperature of the air, and they thus lay for a long time. Being terrified by the noise of those who came on them, they wished to fly, but they could not find the entrance of the cavern before they were caught.[3]

The story sounds so much better in its original telling. And the gloriously serendipitous truth is that Ralph of Coggeshall wrote those words just down the road from where the Honywood Oak lives, at a time when the tree was first finding its roots in the earth.

Of course, there is also that other famous instance of a Green Child that I had stumbled upon some time back – the case of Peter the wild boy who had been found in the forests of Germany in 1725. He, too, has already briefly appeared in these pages. Yet, like the Green Children of Woolpit, I cannot seem to shake him from my thoughts. If not actually green in colour, he had certainly been a child of the woods. Daniel Defoe wrote of the boy that he was 'like a Creature abandoned by Nature itself . . . perfectly wild, uninstructed, unform'd' and that 'he was found, or, as they stile it, was catch'd in a Wood or Forest about Hamelen in Germany'.[4] After he was brought to London, distinguished guests of King George I came to see the poor boy, dressed now in smart clothes, a freak of nature to be gazed upon and amazed by, until inevitably interest waned. Yet the wild boy lived on. He turned from forest child to Georgian man. When he died aged seventy or so, he had a full white beard and could answer a number of questions though only ever with the same two words. When asked what he was, he had been taught to answer, 'Wild man'.[5]

Enrapt in fireside reading of wild or Green Children, I think again of the creature that flickered past me in the dawn light. I really had seen it. Even now, I can see the blur of green, the long hair, and can hear the sound of footsteps running down the road. Whatever it was, it was real to me.

13 March

On 26 July 1852, Henry David Thoreau wrote in his journal: 'By my intimacy with nature I find myself withdrawn from man. My interest in the sun and the moon, in the morning and the evening, compels me to solitude.'[6]

I have found the same is true. As you spend more of your time with oaks, that same withdrawal process from man takes place. It is an inevitable consequence as trees become more commonplace companions. Yet it is in no way disaffecting but rather the opposite – for as you spend more and more of the day, from dawn to dusk and on into night, in the company of oaks, so you feel more at ease with the world and so, too, does this life seem to gain a sense of meaning. You rise with the dawn and you watch the sunset. You gaze on the moon and the stars at night and then sleep deeply so as to get up with the sun once more the next day.

Virginia Woolf's Orlando was similarly withdrawn from other humans by his intimacy with nature.

Orlando naturally loved solitary places, vast views, and to feel himself for ever and ever and ever alone.

So, after a long silence, 'I am alone,' he breathed at last, opening his lips for the first time in this record. He had walked very quickly uphill through ferns and hawthorn bushes, startling deer and wild birds, to a place crowned by a single oak tree.

This morning, dawn is delayed again by mists and fog.

Two hours on, the sun is steadily winning – streaming through the veil. I take coffee and sit beside the Field Oak and in time a goldfinch sings from one of the outer branches. I watch a blue tit hunting in the heights of the oak and think of those tiny green tortrix larvae and wonder if they have already started to emerge or whether by some sense beyond our knowing they are able to wait until the moment the buds begin to open. For now, the buds are still brown. A robin and a wren sing their separate melodies. The blue tit has gained a long tail – transformed in the heights of the oak.

I head away to Two Oak Hill, down past the church along the field edge, listening to skylarks while fieldfare scatter from the hedgerow. A thrush flies and writhes and turns with an ethereal pale beauty. I think of how Orlando has his hill 'crowned by a single oak'. I have Two Oak Hill.

Woolf tells how Orlando came to his oak hill in summer:

He sighed profoundly, and flung himself – there was a passion in his movements which deserved the word – on the earth at the foot of the oak tree. He loved, beneath all this summer transiency, to feel the earth's spine beneath him; for such he took the hard root of the oak tree to be . . . gradually the flutter in and about him stilled itself; the little leaves hung, the deer stopped; the pale summer clouds stayed; his limbs grew heavy on the ground; and he lay so still that by degrees the deer stepped nearer and the rooks wheeled round him and the swallows dipped and circled and the dragonflies shot past, as if all the fertility and amorous activity of a summer's evening were woven web-like about his body.[7]

On Two Oak Hill, there is a distinct feeling of winter. I halt and lean against the furthest oak, hidden from the winds in the lee, feeling the tree, 'the earth's spine', and peer down to the woodland and the stream. A solitary magpie flies over, high and heading north. I wonder at such auspice, shelter and sip coffee. Two humans walk beside a dog, beneath the hill, four hundred yards or so to the south.

I tuck myself down.

4 April
In the oak we know no fear.

By the Field Oak at dusk: the birdsong is already up; the last rays of sunlight are ribbing the western side of the oak. The fire I lit an hour or so earlier in a far corner of the field still burns and the smell of wood smoke lingers in the air though rain has come to dampen the flames, nearly dousing them entirely. There is a wonderful unity of elements: wood, air, fire, water.

The rain comes again as the sun dips, bringing the tinny echo of raindrops on the oak leaves. The dying of the day seems to come quickly tonight. The wren, lost against last year's leaf in the gully, is a rattle of sound and a brown flicker. The fire must be nearly out. I peer south and see ice-blue smoke still lifts against the hawthorn, rising like djinn. I rise, too, to head in. The oaks have turned to black silhouettes against the dusk sky.

5 April
Winter is an amnesiac.

The long, hibernatory months have pushed the memory of spring's rebirth very deep. Yet life is there. It can be seen in the green blades that rise from the brown soils and turn the bare earth to grass. It is there in the buds of the oak that swell now on the furthest branches. In the city the magnolias will have already flowered with their magnificent grace. The coloured petals will lie on stone floors. The first leaves of the horse chestnut tree are already out above the pavements, unfurling soft and silk-eared from their brown casings. Not so here in these colder climes and starker worlds where the remembrance of freezing nights remains real. In the country, things take time to flourish.

I reach Two Oak Hill. Guns clatter away to the south. I sit in the lee of the root limb. Then, for a change, I walk up to the other oak that sits upon this hillside and whose roots are even more exposed in these sandy soils that wash so easily away with rain. I kneel before the tree and peer right through the mesh of roots. A shaft of sunlight shines into the sandy space formed beneath the tree, littered with empty acorn cups. It is a perfect squirrels' larder – dry and hidden from all.

In German, the word for an oak is *eiche;* an acorn is an *eichel.* An *eichhörnchen* is a squirrel, or, one who stores acorns.

This is a battered and bruised oak, roots unearthed and skin torn. I look up and see the crown of the oak is also tattered – a stag-headed form long ago lost to a gale. Yet like all oaks, whether torn and battered or grandly magnificent in appearance, this oak is a world for many thriving communities of creatures. Whether alive and flourishing or decrepit and dying, each oak is a universe to be inhabited. And I know that the stag-headed heights of the oak are merely the tree adapting to the dry conditions it finds itself in – the sandy soils, the exposed roots – and so is withdrawing within in order to save energy, stripping back to its essence. The jagged, lightning bolt of the stag-head does not mean the oak is dying. Rather it means the oak is reacting to its environment, striving to survive, to live.

I walk back and linger in the graveyard of the church a while as skylarks sing above me. There is another strange sound, too, as of a radio tuning. And I know in an instant that it is the call of a peewit, though it takes a while longer to find those round wings flapping and whiffling. It is indeed a peewit. And yet the happiness which at first I felt at hearing its cry soon fades. Only a generation ago, these fields were full of such sounds. Now I follow the flight of that solitary bird, see it standing alone on the soil, and the sight of the poor creature makes me want to weep. How long before there are none? How long before the sound only exists as a recording on a technological device long ago considered archaic?

A fresh grave has been dug in the quarter of the field recently consecrated to form a new fragment of the ancient graveyard. Even here, more human bodies fill the land. And all the while, as the world gluts with humans, the rest of nature squeezes into ever smaller portions, ever more impoverished patches of this earth.

10 April

A Sunday in early spring.

In the afternoon, I weave along a rabbit trail through a meadow and halt before a young *Bombus terrestris*, a vast bumblebee which rests in the shelter offered by the fresh grass. When the bee flies, I walk on, up to the heights of Two Oak Hill and there realise for the first time that I can walk even further on, straight up and into the second oak, climbing with ease up the branches like a ladder into the sky until I am twelve feet from the ground. I sit and say to myself that here would be a perfect place to be in a few weeks' time when the oak buds are fully open and the sun shines and the blue tits gather green caterpillars for their young broods.

I stay there a while, one hand clasped tight around an arm of the oak until I can feel how the heat from my hand has warmed the wood of the oak. And I gaze across the lands beneath me to the south from where I have walked.

So many writers tell of the wonder of sitting beneath an old oak tree, but to sit *within* an old oak tree is another thing entirely. You step away from being your human self to becoming someone else – some form of oak being.

I sit upon this oak bough. Raised from the ground, I sense the day transforming from the dull drudgery of human needs that have filled the morning to now, when a pure immediacy exists in simply sitting in this oak tree. I feel the roughness of the oak skin on mine. My fluttering thoughts vanish as I climb. There is something utterly liberating in being within the oak. To rise a dozen feet or so from earth entirely alters perspective. Vantage transforms viewpoint. My right hand holds tight the rough skin of one oak limb. I sit upon another. And from here I lean back against the oak trunk and peer about me, feeling perfectly in the present, seeing all as it truly is.

For an hour or so I sit upon that bough. I feel the tree swing and sway until with some reluctance I muster up and begin my descent. Now back on this earth my body feels somewhat lighter, as though the experience, the time passed there upon that oak bough, suspended in the air, has stripped me a little of some mortal weight. So as I walk back along these well-worn roadways, my day does feel different; the burden of life is less heavy. All is lighter. Even as the clouds build behind me and block the sun and turn to grey, the day feels brighter.

15 April
Drizzle falls.
I return to sit in the familiar western nook of the Honywood Oak. All morning rain has spoilt the illusion of spring. Yesterday, the roll and clap of thunderstorms brought sharp April showers, so fresh and real. Today, the grey of winter returns. A peacock calls from the north. A chiffchaff, newly arrived, keeps up a merry song. Yesterday as I walked into town, I disturbed two little owls that for some reason were awake and alert in the afternoon and sitting on an oak. One turned and stared angrily before it flew away. The other plunged a moment later into a maze of trees in a flurry of feathers. It is true that we humans are seen as killers stalking the land. Even if our intent is innocent, all other living things around us believe we are there to kill. The pigeons burst from their perches in the highest branches of the trees from a hundred yards away as I approach even though I have never fired a shot at them. They still fear. So they fly anyway. All flee before us. Those that can fly, fly; those without wings seek shelter. As any human approaches, the rabbits on the hillside will tear for the dark safety of their burrows; the deer will

plunge for the cover of the woodland. All creatures have learned to run before us.

Today, the tree looks as I feel, a little forlorn and damp, sodden in this sudden return to winter weather. In rain, the wildlife vanishes. The birds that flew in yesterday's sunshine have gone. All is silent. All signs of existence have been extinguished. Humans, too, tuck down and hide inside on days like these of endless drizzle.

And yet, what's this?

It is a treecreeper that flies in from the south and scurries about the bark, setting to work its way around the high boughs of the oak. And another arrives. And the two work together, as though turning circles of some invisible thread about the heights of the vast northern branches. And then one vanishes into the dark slice of shadow between the dark trunk and the pale limb.

I am held in frozen glee. I feel the moment, ecstatic in stillness. From the ground I stare up into that distant darkness for a few moments but the bird does not re-appear. I think of the treecreeper forming a nest beneath the fold of old oak skin – dry and sheltered and soon to become a home to its young.

17 April

The day fades.

I reach the oak just as the sun sinks. On Two Oak Hill, I climb and sit on the secure seat of this bough of oak. My back rests against the trunk. My eyes turn to the western horizon where streaks of mauves and pinks signal the glorious dregs of day. As I fled the city a few hours back, the promise of this moment drove me on. The oak is warm after a day in the sunshine. For a few precious moments all seems well. And then the cries of the dusk chorus are broken by the sombre chime of the church bells from over the valley. A pale night bird flies in the gulley below. I hear the distant voices of people and I do not know whether to stay here in the oak and try to hide or to slip back to earth. I think of Cosimo, Italo Calvino's baron, who climbs into a tree one day and stays forever in the realm of branches and leaves.[8]

I descend – down from air to earth. The cold of night seeps instantly in as the sun slips beneath the distant hillside. My feet touch the ground and I scurry away through the meadow grass.

18 April

I wake before dawn. The small birds about me wake, too. We rise together. I stir and move and scatter sparrows.

And in that dawn light there appears a delightful surprise.

'*Hirundo domestica*,' I whisper to the spring air.

It is the first swallow of the year and the sight of that bird fills me brimful with emotion.

20 April

I step out towards Two Oak Hill. A single swallow turns above my head, tail sticks wide apart, searing through blue sky. Sunshine has transformed the world today. Two pairs of crows whiffle down towards earth over still un-ploughed fields. I walk south.

I pass an old lady who is cutting the grass in her garden with a pair of scissors. She is startled when I speak, though I only say, 'Good morning.'

'Hello,' she replies, but she is hardly there. Her hair is ashen white and thin. Her eyes are dewy and wide.

I walk on, unsettled, on to the meadowlands, following the rabbit tracks that lead up to Two Oak Hill. I climb the oak and secure my seat, my back against the trunk and stay there, hidden and sheltered from the eastern breeze that blows. I listen to the sound of the wind in the tree around me, the calls of birds, and marvel once again at the immediacy of the sensation of sitting in that oak, held from earth by the arms of the tree.

I look about me and see that no more than half a dozen buds have opened to reveal their waxy, shiny store of leaf. It is the same for the oak beside me.

Time passes.

A dog barking from the farm wakes me though I have not fallen asleep.

26 April

This week has returned us to winter. The wind is so cold as to chill my hands though sun shines and skylarks sing above me.

I thought last night of those vast oak footprints that Jonathan showed me some years back, all that remains of the great oak forest that once stood across these lands. And I remembered the feeling of loss I felt then was as though I was walking across a silent battlefield.

If we are not aware and alert, other men will come with chainsaws and take down the remaining bodies of the great oaks that live upon this earth. And if we do not protect the young oaks that have only now begun to grow, there will be no great oaks in the future. In eight hundred years' time, will there be any ancient oaks in these lands?

1 May

I am filled to the brim with emotion again today and fear I cannot stop it overspilling.

Nature mediates. I am eased, as ever, by the simple act of stepping out, of walking and seeing, observing all before me. Beside the Field Oak, I watch the interplay of birds and tree – the emergent leaf now a lemon-coloured coating above me – and realise with a certain sadness how the human is always outside that interplay, that mesh of life. Was there a time in some ancient prehistoric world when the creatures did not flee before us, the birds did not fly from the branches of the oak as we approached? Was there a time when humans did not strike fear and alarm into the natural world around them?

We are of nature. Yet we see ourselves as separate.

Beside the Field Oak, I sit beneath the garbled warbling of a garden warbler whose mate's reply I hear in the still between songs. The day is so much more alive now with birdsong, the pipes and squeaks from one to another in that realm above. And then there is the full-fluted chorus of song at start and end of day.

The leaf canopy steadily fills the space between the boughs where sky sat all winter. Leaf fills space. The oak expands. The tree breathes and with each breath, it grows – working towards those qualities so often given by people of this land, of strength and valour and wisdom: the wise old oak.

I strive to spend all the hours I can in the company of oaks, and yearn for some transference of those qualities until I may hold something of their essence, their character, too. Those qualities which we ascribe to this oak beside me are ones we all would wish for. I think of Thomas Hardy's gentle hero Gabriel Oak from the novel *Far from the Madding Crowd*, who even when fate is cruellest finds within him 'a dignified calm'.[9] It is in the name.

We, as individual human beings, like Gabriel Oak, can seek to strengthen our own sense of worth, I think. We can be stronger, fairer on this earth, and help to home and house and care for the creatures of this world that live cheek by jowl beside us.

With a torturous forbidding cry, a jay flies in from the north and perches on an upper bough of the Field Oak and from there squawks and then flees east in the company of three other jays whose flashing wings I see before they vanish into the copse of willow next door.

The peace that is restored lasts only seconds before their rancour is recalled again, as the four fly north towards the wood where buzzards lie.

13 May

Twelve days have passed since I sat beside an oak. In winter, twelve days can pass and nothing appears to have changed, but these are twelve days in May. In that time it seems the entire world has changed. The oak leaf that was shyly peering forth now boldly fills the sky. The birds that then I could watch as they hopped from branch to branch, now sing unseen, hidden beneath that canopy of leaf that in the bare bones of winter felt as though it might never grow again.

Spring is here once more. Twelve days in May and the world transforms. All England turns green. The earth that seemed so dull and dead through those first long months of the year now pours forth life.

I hack with an old, dead branch through the undergrowth of bramble and nettle to reach the hidden camping chair in that space beside the Field Oak. A whole bank of green has grown up – nettle, bramble, cow parsley. I am enclosed in an emerald cave. Above me, the young oak leaves are out and grow ever wider, their skin hardening to the air though still waxy to the touch. I stare at the oak. The breeze is up. The sun shines, casting a lemon-yellow light through the canopy. In all its days, this oak has never been so alive.

In Leo Tolstoy's *War and Peace*, an oak tree is a symbol of vital significance for Prince Andrei, who grieves the death of his young wife, Lise. He notices the oak first at the start of spring in 1809 as he visits a country estate in Ryazan a hundred miles from Moscow. The birches and the alder and the cherry trees are all touched with fresh green growth. Unlike those around him, Andrei feels nothing. He sees an oak,

> With its huge ungainly limbs . . . its gnarled hands and fingers, it stood an aged, stern, and scornful monster among the smiling birch-trees.

The oak, like him, 'refused to yield to the charm of spring'. As he travels on through the forest, Andrei turns back to the oak several times 'as if expecting something from it'.

> 'Yes, the oak is right, a thousand times right,' thought Prince Andrei. 'Let others – the young – yield afresh to that fraud, but we know our life is finished!'

The oak and Andrei are united:

A whole sequence of new thoughts, hopeless but mournfully pleasant, rose in his soul in connexion with that tree.

It is six weeks later when Andrei returns to the same forest. He is alone and leaving for home. The night before he has heard the ecstatic voice of young Natasha as she stared out at the starlit night, delighting in the wonder of all. Andrei seeks that same old oak.

'Yes, here in this forest was that oak with which I agreed,' thought Prince Andrei. 'But where is it?' . . . and without recognising it he looked with admiration at the very oak he sought. The old oak, quite transfigured, spreading out a canopy of sappy dark-green foliage, stood rapt and slightly trembling in the rays of the evening sun. Neither gnarled fingers nor old scars nor old doubts and sorrows were any of them in evidence now.

It is a moment of revelation, an epiphany.

'Yes, it is the same oak,' thought Prince Andrei, and all at once he was seized by an unreasoning spring-time feeling of joy and renewal. All the best moments of his life suddenly rose to his memory . . .

He is seized with a firm belief in the future.

Everyone must know me, so that my life may not be lived for myself alone while others live so far apart from it, but so that it may be reflected in them all, and they and I may live in harmony.[10]

His grief, despair and despondence at life are gone.

16 May

The grass is thick and lush beneath my feet. I walk out through the wooden gate and onto the pathway that leads to the Honywood Oak. There is the familiar honking of the greylag geese. I see the Honywood Oak as though for the first time. The oak is in bloom, held in a green cloud of leaf.

It feels as though it was another world, another lifetime, when I came here before. I smile to myself, renewed. I am exhilarated.

I had forgotten how wonderful it is to sit here upon this bench and observe the day of the oak unfold, the retinue of animals, birds and insects that come and go and who form the community to which the oak is central – a single living entity whose frame and structure, whose being, is so vital to so many others. And I know that I am merely another creature who comes and steps within the shadow of the Honywood Oak, who rests a while against the comforting trunk. I know that I am just another individual to whom the oak gives comfort and solace.

I rise.

I step from the bench to sit a while beside the oak.

Beside the Honywood Oak, my left hand leaning against the bark of the trunk, there is a gentle flap of wings, a flitter – a treecreeper lands some six feet above me. That delicate curve of its beak is so close. I cannot breathe. The bird is holding a spiky mass of fly. It pauses for an instant, then climbs up the main bough towards a tear in the trunk and vanishes into the oak.

It seems that as I become more withdrawn from my fellow man so the creatures of the world no longer fear my presence.

23 May

The day starts grey. Sunshine entices birdsong. Beside the Honywood Oak, a chiffchaff calls incessantly from the top of an ash tree. Each *chiff* and *chaff* are mirrored in the lifting of the bird's tail feathers. I see him now turning this way and that, seemingly as agitated as the calls suggest. Perhaps he feels that time is ticking on. The season is turning. Still he has not found a mate to answer his call.

A treecreeper appears and works a threaded path up the western edge of the Honywood Oak. The pale underside of the bird is perfectly matched as it steps over a section of the oak where the bark has peeled back to reveal the exposed sapwood. I watch as it digs beneath the edge of the bark, the curved beak like that of an acorn weevil, able to reach so far within the oak. It delves and pecks and probes further, flicking fragments of loosened bark twenty feet down to the earthen floor. A second treecreeper appears, also on that western front of the oak, with something already held in its beak. I can see the splay of insect legs and wings. It flits and flies and is gone to one of the outer branches. I close in. I creep across the leaf litter until I feel the rough touch of the oak bark on my skin. They are both gone from sight. Something stirs. The north wind blows in. I feel a thread of silk gossamer on my hand and gaze upwards into the oak and in a flick and a whisk the two treecreepers are back as though from nowhere. I step back. I will watch and wait a while. The birds meander upwards, winding their way along the outer edge of the oak. One flits to a branch and suddenly I lose them both as they go their separate ways skywards. One appears again and alights on the embossed scar of a horse-shaped pattern of greyer skin where hundreds of years before a limb once grew and then fell. There the treecreeper vanishes. It has tucked itself beneath a flap of bark: an entranceway within the oak where young treecreepers wait to be fed. All this I see and learn by simply standing back and waiting and watching the ways of the Honywood Oak.

I creep closer and stand upon the seat that is my sheltering space in the western quarter of the oak. And I peer from two feet away into the entranceway beneath the bark. One of the treecreepers is inside, the other lands only three feet away. I see the speckled coat, the curved beak, the bent wings – and the sprawled legs of the fly held there.

A robin appears whose red breast is so clear and bold, a splash of colour even from thirty feet away. It stands beside the dark circle that is the hollow end of a vast trunk that faces west. I think of Robin Goodfellow – fairy, wood sprite, bringer of merriment. And the robin sings.

It seems that the sound of that song is the embodiment of the spirit of the tree, is indeed the voice of the dryad, the song of the oak.

I am lightened.

I walk south from the Honywood Oak and take a moment to see that other grand oak nearby that remains from the ancient woodland cut down half a century or so ago. It, too, is a remarkable tree. Five hundred years old or so, its inners are blackened by a lightning strike. It is known as the Screaming Oak from the figure of a horrified face on the pale sapwood; the eyes are two dark circles that once were homes to birds. The torn, dishevelled nature of the tree draws forth feelings of sadness that such a great oak can be so maimed. And yet, there is a canopy which, though stunted, is still verdant with fresh leaves and the tree remains alive and may well do for many more years.

It is good to be reminded that each individual oak that I get to know has some uniqueness portrayed in the physical characteristics of the tree. I pace and step about the earth around the Screaming Oak and see the specific facets and mannerisms of the tree: the two vast torn-off limbs to the north; the wild bees that stream from the entrance to another dark hole made many years before by the hammering of a woodpecker. That hole has been a home, a nest from which so many songbirds have fled into the light over the years. And now it is home to a colony of bees.

I walk back north to the Honywood Oak, a hundred yards away, the only other ancient oak still standing here. And I vow that I will try to feel the distinctiveness of each individual living being that I meet.

When I finally get to meet the artist Stephen Taylor, I'm rather late. I blame Sunday drivers. We have spoken on the phone and found a window in our Sunday afternoons so I drive slowly over to Ely, through sunny Suffolk lanes until the skies open ever wider in the fenlands and a sunlit cathedral sits on the horizon.

At his house, I'm led up the stairs to a living room-cum-studio where a painting of a waterfall still on its easel dominates the space. He has just finished it. The water that both rushes in torrents and mills in pools is incredibly lifelike. Stephen is a remarkable painter of nature. The detail and verisimilitude are astonishing. Yet Stephen knows that it is the oak paintings I have come to talk about. A decade or so ago, he spent three years creating fifty paintings of the same oak tree in a field in North Essex.

He stands in the kitchen and makes tea and talks of

a time when he spent days and nights throughout the seasons painting a series of portraits of an oak tree that gained titles like *Oak with Crows, Oak after Snow, Oak at Night in Winter*. He appears as a gentle figure in his comfortable brown fleece and blue jeans, yet there is an earnest intensity in those eyes behind his glasses. It is there, too, in his words. He is ardently passionate about his art and about telling the story of his work.

We step into an adjoining room where a large painting of a wheat field fills one wall. Stephen starts to tell his story.

'West Bergholt is there,' he gestures. 'Ronnie Blythe's house is two miles this way.' It is a field in North Essex.

'And there's your oak,' I point out.

'It is indeed,' he says.

Only, at that time, Stephen wasn't specifically looking at oaks. He had been invited to visit an eight-hundred-acre farm owned by two friends who had returned from New Zealand. Stephen asked if he could paint on the farm and when they readily agreed that is what he set out to do. Other than the farm manager, Bill, and the farmhand, Reg, Stephen knew that there was no one else he would run into. He had free rein. It was a big area to stroll in, and having the liberty to wander about such a space proved hugely therapeutic.

'I was knocked about a bit psychologically when I started all this,' he states. 'Being in a place where every-thing was green and it was your field, so to speak . . . it was a bit like outdoors therapy.' He laughs a little, then slows and speaks each word precisely. 'It allayed anxieties.'

I murmur appreciation but let him go on.

'I absolutely felt that it was the grounding that I needed. And that let me paint more precisely.'

I knew something of the tale already – remembered parts from what Stephen had written in his book *Oak*.

'You'd recently lost both of your parents?' I say.

'And my girlfriend and my best friend,' adds Stephen.

Within some six years, he had been stripped of all those dearest to him.

'You have this intense grieving, and then you get another loss, and then you get another one . . .' he says in a flurry. 'And you feel you're in limbo. You're not attached to the everyday world.'

'And the farm became some kind of safe space,' I offer.

Stephen laughs softly.

'Yeah,' he agrees. 'A safe space.'

At first, it had been the wide-open landscapes of the farm that provided some kind of solace. Returning time after time allowed an immersion into that place.

'It was a grounding process,' Stephen says. 'You see more and more. Then you forget yourself. Then you find out who you are.'

He smiles. I look back to the cornfield beside us.

'It's a weird process,' he adds. 'It sounds a bit weird, doesn't it? But the more I was there, the more I realised that nature is just a primal source. You can keep going back. Life was chaotic but you got the confidence to deal with the chaos.'

The fields of the farm were a beneficial environment. Being out there Stephen could return to his professional working practice – painting. And the act of painting provided the therapeutic activity he needed. He could focus on the task before him. He concentrated all thought on painting one particular field that stood on a slight rise, the world rolling gently away before him – the result was the picture that hung on the wall in front of us. When that was complete, he had simply started to paint another.

We walk back through to the living room where a familiar cornfield landscape fills the wall. It is another painting of the same field. Two pigeons fly through an otherwise empty summer sky. A lone human figure walks a tractor furrow.

'There's Bill, again,' points out Stephen. Bill is the farm manager.

It was an episode in New York that made him see a deeper, personal truth to what he was doing. The painting now before us had been on display in a vast exhibition space in Manhattan.

'A gangster from the Bronx walked over. He had these brown and white shoes on, wore an extremely expensive camel-coloured coat,' Stephen laughs in remembrance. 'I mean, he almost had spats on. Some big guys behind him. And he walked by and I button-holed him – politely, very politely – and he realised it wasn't a photograph, and he said, "Holy Moley!"' Stephen mimics a New York drawl.

'And then he said, "Is that your dad?"' Stephen leans over and points to the lone figure of Bill in the painting on the wall. For a broken second, he falls silent.

'Oh, right,' I murmur.

Stephen hadn't thought of it at the time but there was something distinctly poignant in that question.

'Bill became a kind of . . .' His words drift away. 'I didn't think of it at all.'

'You think there's an element of that . . .' I start to suggest.

'Yeah,' agrees Stephen. 'Bill's a working man walking through the farm. My dad died a few years earlier. He was a working man. You know what I mean?'

I do.

'So that strange New York encounter opened me up to the psychological dimension of what I was doing,' he says. 'I was telling myself it was exploratory nature realism. But it was something else, too. I was really taken aback by that American guy saying, "Is that your dad?" I mean it must be! And I never realised. So there's certainly sublimation.'

'And that feeds into the art,' I add.

'It does.'

After exhibiting his landscape paintings of that cornfield, Stephen found himself back on the urban streets of Colchester. He returned to those eight hundred acres of farmland merely to enjoy the summer months in the freedom of the field he had got to know through painting.

'I was just hanging out in the field – bird-watching and chatting with Bill,' he says. 'I was waiting for a subject to pop up.'

Then he found the oak.

He painted his first oak tree painting, *Green Fire*. When he painted his second image of the oak it was in the autumn. The oak had transformed. Golden, bronze and orange oak leaves filled the frame. Stephen began to feel something shift as he held both paintings in his hands and saw the variance. It was the same oak painted in the same way, yet so different. Something stirred within him.

'There was some kind of raw feeling . . .' Stephen's face scrunched up as both hands turned to claws before him. 'I was like, "F . . . ing hell",' he says.

I startle somewhat at the words. But I know what Stephen means. We have moved to that territory where there are not quite the words to describe the feeling experienced.

'That's the moment,' I offer.

'That's when you know.'

Over the following three years, one single, two-hundred-and-fifty-year-old oak became the sole subject matter for Stephen's painting. Those fifty portraits that captured the oak in every kind of weather and season lent the tree an individual significance. Rarely does a single human receive such attention from a painter, let alone a tree. Each portrait shows the oak in a distinct aspect of being. In *No Moonlight*, the oak is a blur of darkness against a darker line of hedge. In *Spring Haze*, the oak is alive with the first flush of the new growth.

The oak becomes the central, static feature around which all else revolves. In his notes to the paintings in the book *Oak*, Stephen tells how with one work:

> When I was painting this, at the back of my mind was 'Burnt Norton', a poem from *Four Quartets* (1944) by T.S. Eliot that associates a tree with 'the still point of the turning world'.[11]

In another, a pigeon tears across the canvas, over the green baize of oak leaf. I take Stephen back to that image of the pigeon in the oak.

'I love it,' he says with a growing smile. 'I love the bird passing through.'

'It's that instant captured,' I say.

Stephen nods. He is all enthused again, bravely seeking to capture in words feelings that are indescribable.

'The bird is . . .' he begins. 'Like in Gerard Manley Hopkins . . . the bird is a kind of secular holy ghost, isn't it?'

'Absolutely,' I agree.

Stephen muses on exactly what it was he was attempting to capture in that slicing vision of the bird and the oak, the intensity of a fraction of life.

'It's a moment of sight, isn't it,' he explains. 'And gone as soon as it's happened.'

I mention the Japanese Buddhist term of *satori*.

'It's exactly that kind of moment. A bird tearing past. A hare halted and staring. A fleeting glance into the true nature of all things.'

Stephen turns us back to T.S. Eliot.

'That's the same glimpse of the bird in *Four Quartets*. "Quick, said the bird".' [12]

As a teenager, he had carried Eliot's book all round France.

'That's my sacred text,' he says. 'It's been in my bag since I was eighteen.' We have drifted into matters transcendental. It is easily enough done. When we get back to the oak it is to that moment in spring when the oak comes alive once more.

'I saw the oak buds come out,' Stephen says. 'And they're really weird – they're pink, almost like plasma, like blood. Very peculiar. And if the light hits and it's a sunny day, they're a kind of warm salmon colour. Like life.'

He is grinning at the memory. He makes another literary reference – to Prince Andrei and that oak which springs to life in *War and Peace*, reassuring Andrei there is a happy, harmonious future for him.

'That was another reason I turned to the oak. You know how the oak in winter is almost like a rock – black, static in the land. Ancient. Dead. And then you suddenly get these little pink things coming out by the thousand in the spring and it's that emotion of "Wow".'

I know.

'And it strikes me that it's something to do with the

longevity of the oak that has an important cultural significance to us,' I say. 'It keeps coming back.'

'It does,' Stephen agrees. 'It's out of human time. You can't make the leap to the life of an oak. It's a kingdom leap. You can't do it.'

'And we have some kind of innate respect for that. I think it provides us with this comfort. That an oak sees through generations of us. So perhaps it's a tree that we go to with grief and it does some sort of easing of that grief.' I search for the right way of putting it. 'Even though we're not recognising that at the time. We're not deliberately saying, "I feel sad. I'll go to the oak." But on another level of consciousness, we are eased. The oak is always there. We can always return to the oak.'

'Yes,' Stephen says.

'I mean, you spent three years with that oak.' I point out the oak in the painted landscape on the wall before us.

'Yes.'

'You know that individual tree so well. You know the individuality of that tree,' I say. 'You could pick that oak out in a line-up of a hundred oaks.'

Stephen laughs.

'I could. I could pick it out of a line of rogue oaks in the police station.'

It is quite an image.

'That's the one that fell on the child!' He jokes in the voice of an accusing court barrister. 'And it meant to!'

We both laugh.

But I am intrigued to know how it felt for Stephen

to return to that same oak each time to start another portrait.

'It felt like going back into a room in my own house. For a lot of the smaller portraits of the oak, I had exactly the same spot, sitting in the field. So the bit on the ground got worn hard where I had sat. Over three years, I went back to that place. I had a relationship with that place. It's a bit like a comfy chair in a favourite room. I kept going back there.'

On one level, Stephen kept going back to the field and to the oak in order to carry out the 'pedestrian craft' of painting. It was that which acted as the ballast of his life. Some time later, after the exhibition of his fifty oak paintings, he went back to the field once more.

'I lay at the foot of the oak tree on the verge of the field and it felt like a sofa. The tree was beside me. The sun shone. There was this wonderful skyscape. And I had one of those lovely moments. It was just so very, very beautiful.'

I think of Orlando on the ground beside his oak tree.

Stephen knew that he could either keep coming back to that place for the rest of his life or he could move on.

'And the door shut. A lot of water had gone under the bridge by then,' he explains. 'All the reasons for being in the field, for being by the oak, had gone. The grieving had long since finished.'

He had spent seven years going to that field, three of which he'd passed at the oak.

'You weren't deliberately going there to be healed,' I say. 'And yet in some way . . .'

'It was helpful,' Stephen adds and pauses and then repeats. 'It was helpful.'

'And the oak became part of that process.'

'That's right.'

24 May

I head for Two Oak Hill and from five hundred yards away see the two oaks framed on the horizon, rising green. Each has a quite distinct hue. The furthest, in whose lee I have so often sat, is a paler, lime green; the oak into which in recent times I have stepped up that bough ladder, has a far deeper green touch to its canopy, though why that should be so I do not know.

In merely a few days the meadow has been transformed. The rabbit track I follow is now deeply indented in the meadow grasses. Clumps of buttercups now paint the grassland with splashes of colour. It is the way with spring. The reappearance of life on earth brings such hues of colour – and we realise how monochrome winter really is.

I startle a skylark – and another – which start from the thick grasses and curl round in a wild parabola as though tied together until they are back on the ground and close to where the nest they are building is neatly hidden.

The first of the oaks is now transformed even more into a mushroom of green leaf, thick and low, for these two oaks are exposed here on this hillside so both tuck down close to the ground. This oak, into whose shelter I will soon climb, is marked with a two-pronged stag-head of grey branch which strikes out from the roof of the canopy and appears like an antler or a fork of lightning. It adds such a distinct look and sense of personality to the tree. The canopy of the lower oak is far more fragmentary. The paler leaves that I could see from half a mile away do not form a complete roof to the tree, but rather a tattered and incomplete shade, a smattering of cover.

Shakespeare had played on that stag-headed aspect of the oak when he spoke of the figure of Herne the Hunter:

There is an old tale goes that Herne the Hunter,
Sometime a keeper here in Windsor Forest,
Doth all the winter-time, at still midnight,
Walk round about an oak, with great ragg'd horns[13]

The stag horns stand on the head of the ghostly Green Man figure of Herne but it is the connection to the stag head of certain oaks that strikes me. The stag head unites the deer of the forest with the oaks.

I am reminded of those ancient antler headdresses unearthed at Star Carr in North Yorkshire that date back some 10,000 years to the Mesolithic period. Around two hundred such masks have been found, each forged from the antlers of a red deer and apparently designed to be tied around the head for some symbolic purpose. The human becomes the stag. The stag-headed aspect draws together the human to the deer and to the oak. As Stone Age shamans wore the antler headdresses of Star Carr to become the red deer, so, too, perhaps they were taking on something of the forest – becoming, with their stag-headed mask, part ancient oak, part deer.

1 June
Greyness fills the air.

Earlier, fog had shrouded the landscape. The land was held in suspension. There was a tangible notion that this grey blanket would soon rise and reveal a glorious other world.

I thought of a dawn when I had awoken on a Scottish island to find myself encased in a dense sea fret, and as I had passed the moments preparing myself for the day, had felt that self-same sense of otherworldliness – of another realm entirely being a fingertip away. I had walked through the wild flower dunes, the machair, of that land as though half-blinded, or stumbling through a lunar landscape at dusk. The sun burst through, scattering the sea fret and I found myself standing in the stone enclosure of a Bronze Age house exactly as the sunlight found its way onto the earth.

Now I listen to the fair sounds of skylark song on this so foul a day. Though there are still three hours or so before the sun is due to set, a sense of twilight fills the air. The sun is not about to burst through. Light will slowly fade over the next few hours. The greyness will find a darker hue and the day will steadily turn to night.

The wind rises. I walk on. I pass the leafless limbs of the boundary oak, which seems to have suddenly died. Crows are working their way over muddied and puddled soils. It is as though the seasons are reversing; as though time is turning backwards and the earth is revolving through spring to winter. Blustery bursts of cold wind blow through the gaps in the hedge. Cow parsley on the field edges shakes wildly. A swallow, alone and dark in the grey sky, looks utterly lost in such weather. The call of a chiffchaff comes through with each gust of wind.

I head down the hill past the house of that old lady whom I had seen snipping at the grass in her front garden with scissors earlier in the year. Some good soul has mown her lawn. I feel gladdened at the sight. I go the long way to the oaks, sidestepping the sodden grasses of the meadow, slipping through the broken, barred gate to walk the high edge of the hill. The stag-head antler of the first oak is pale in the distance, framed in its green canopy. It is forked like a frozen bolt of lightning. Rabbits fly. I reach the summit of the hill to find a rabbit carcass on the ground before the oak. It is bloodied and crow-torn and it stinks.

I climb up into the oak and find a strange stillness. The tree protects me, hides me from the worst of the cold winds. Its world encloses me. I hold tight to the bough.

On the skin of my fingers I feel the dampness that has seeped into the skin of the oak. In time, I feel, too, a gradual warmth. For some minutes I remain, held by the strange comfort I seem to gain by just being there, and by a reluctance to step back down to earth.

6 June

I walk through the wooden gate to a vaulting blue sky. The goslings have grown so much since I was last here. The greylags arise and walk ponderously over to the lake. I step past them, up the easy incline to the Honywood Oak, which I have never seen looking so well.

The chiffchaff is there again on the top of a pine. Half a dozen long-tailed tits with their high-pitched chatter flitter in the same high branches. A green woodpecker leaves the oak to the north, followed soon after by a sparrowhawk which flies east with a gentle, unhurried ease. It halts on the tufted upper arm of the tall cypress tree on the other side of the stream. As I walk to the oak I feel its eyes on me.

It is the treecreepers I have come to see today. I woke in the night with some strange distorted version of a treecreeper peering at me from a dream and I returned to sleep uneasy. Yet that vision has brought me here today – intrigued and a little entranced by the birds. Somewhere, up in the thickest canopy of the oak, a greenfinch calls with all the glory of summer. For some time, I stand in the lee to the south-west of the oak awaiting the arrival of the treecreepers to their nest, yet that moment never comes. It seems the young have fledged. This quarter of the oak is now quiet.

When I walk away from the oak, the sparrowhawk appears beside me, cruising along with cold elegance, that same air that you imagine a great white shark exudes. It is a sense of invincibility, born of killing others without care or compassion. The hawk had come out of the oak canopy in that upper north-eastern quarter where the calls of greenfinch had been too enticing to ignore.

The day will only grow clearer and brighter and warmer. The lives of those small birds in the oak will be lived today with a glorious intensity. That sparrowhawk will turn and return many times today. Life and death will be all about the oak today.

8 June

I step into the shelter of the western lee of the Honywood Oak – today not from wind or rain but from the sun, which already feels fierce. A hornet, high above, surveys the upper boughs, inspecting the dark circle of an old woodpecker hole. I had forgotten how well my body fits into the oak in this seat – my back framed in the curved form of the oak. To sit here on such a day with little sign of mankind or the modern world is to start to slip gently and happily away – almost from being human. I listen to the birdsong. My own words fall away, for here I have no need of speech.

I imagine somehow climbing into the oak. Simply staying there.

In time my language would fade. Even in mere moments, my world melts away. I start to hear the panoply of summer song in all its layerings and resonances. Just as those wild bees sing their monotonous melody, so, too, it seems from here, beneath and beyond the sounds of warblers, the cries of birds, there lies a carpeting of other sonorous murmurings that we can tune into if we care to try.

I step away and stare once more up into the boughs and wonder at the space and sheer scale of this oak and remember the truth that only two generations back across these lands hundreds of such oak trees lived and would still live now if other men like me had not been told to cut them down. Something of the horror of that act, of the keen sadness of that realisation strikes home into me. If I had been born a hundred years ago, I would not be so drawn to such a great tree as this one. I would not see its size and scale as unique within these lands, for I could have walked a few short yards and found others, brothers and sisters of similar size and scale.

15 June

In the field this morning the swallows are flying four feet from the ground, skimming the tops of the taller grasses and the thistles. I stand by the wooden fence and watch them hurtling towards me and then at the last moment careering away up into the clearing sky. It has rained and the sun is drying the ground of rain and dew. Skylarks sing somewhere to the east. In the west, the oaks that mark the boundary are now a mass of greenery and seem to mirror in substance something of the pale clouds above. I walk towards the oak, the grass soaking my shoes. Two buzzards turn circles in the sky.

Later that day, I walk through the gate that leads to the Honywood Oak and am halted by the soft scent of honey that makes me immediately think of that fragment of honeycomb that sits on my desk at home, which I retrieved from the floor beneath the oak some years before. I step out into the sunshine of the afternoon, to the sight of the heron standing upon the bridge beside the lake, like a gatekeeper. As I approach, it turns its head towards me and lifts from the ground with those wide enveloping wings, heading south and leaving only a gathering of geese before me. They, too, soon depart, stretching and rising and settling on the cold, dark waters of the lake.

The thunderstorm that broke earlier today has left the air cool and fresh. I go and sit a while beside the Honywood Oak. It is not long before I see the sparrowhawk drifting over from the north. He sails on by, past cloud patched with grey, and I lose him.

Beside the oak, a great frenzy of flies and gnats fly –. brought to life and lit up by the late afternoon summer sun. A woodpecker flies in from the north, silent for once. There is that fecund smell that rises from bare earth after rain. I stand on the lee seat leaning against the bark of the oak and breathe in heavily. It smells of earth and leaf and wood and water: of life. Blackbirds call and fly about. A blackcap's fluid song swims in the south. I step back down to earth and look up to see the flight of the wild bees in and out of that dark hole above. The sun is warm against my back. I wander around the oak, peering up and then peering down. Clumps of red heartwood, sections of bark, dried brown leaves and empty cups of acorns, long since carried away. A fresh artist's fungus grows in the west and as I close in on the oak, a pale moth, disguised against the bark, flies off.

I feel a sudden desire to find a way to climb up into that vast wide bowl within the centre of the oak where I have been before, lifted there by ropes. I want to sit up there, in that secure home, to pass the hours and days with little thought but for the signs of wonder that will befall each moment that I sit within the Honywood Oak, immersed in its being, and in the coming and goings of those other creatures which inhabit the oak and call it home.

The matter is so clear, so obvious. How is it that by stepping into the Stag-Headed Oak on Two Oak Hill, or by sitting beside the Honywood Oak or the Field Oak, I feel a calmer and happier person? For some time it seems that I have been skirting around the issue. Yet the central point is that I have experienced a really profound sense of still, of peace, simply by being close to an oak tree.

The problem has been how to explain that feeling. Is it something deeply fundamental to do with a spiritual connection with the tree, with becoming grounded, becoming one with some vital forces of nature? Or is it merely the space and peace such moments offer? Should I be seeking answers from the writings of others on oaks or should I see the experiences as something more personal, more divine?

I turn to science for an answer.

'Why is it, when I, or someone else, sits in an oak tree or by an oak tree, we feel calmer and happier?'

Mike sits back. It is a good opening gambit. Mike Rogerson is a psychologist who looks specifically at the role of the environment in affecting our psychological states. He is also part of a group called the Green Exercise Research Team based at the University of Essex, which looks at the various ways in which outdoor exercise makes a positive impact on us. By some odd tweak of life, I actually taught Mike in secondary school. Now I look across at the neatly bearded face of this young researcher. Time does funny things.

'There are a few things happening,' Mike begins. 'For a start, you're in a place away from your normal world – away from the straight lines that Arnold Wilkins, for example, has done some interesting research into, showing how manmade worlds of straight lines are likely to be more stressful than the curves of the natural world.'

I look about the room. We are sitting in a stale office with straight lines all around us.

'Those straight lines are less likely to be visible from your oak tree so those harder aspects to process visually are removed from your visual field. Then, of course, you're mentally distant from the busy chatter of your busy life. There's a psychological distance whereby you can stop dealing with the immediate, perceived threats – and threats don't have to be someone coming at you with a knife, they can just be situations that we are under pressure to deal with which happen in our personal or work lives – but in that moment in the oak tree, that chatter, those perceptions of needing to do something quickly disappear because physically you're away from those threats and psychologically, you know you're in a different place. All the sounds you can hear, all the sensory bits of information coming in are convincing you that you really are away from that potentially negatively affecting environment that you've come away from.'

It all makes good sense.

In the oak, everything feels more peaceful and calm because it really is more peaceful and calm, and our mind and body know it.

'And there's always going to be that sense of escapism as to why you've come to that tree, as well. People will attribute properties, characteristics to places – they don't have to be places in nature but they often are nature places, surprising or not – and then people think, "I will go there because it will help me relax," and it becomes a self-fulfilling prophecy. You believe that's what it's going to do for you and sure enough, the more you do it, the more it works. Next time you're going to be even more confident that you'll be relaxed.'

Inevitably, I think of those moments when I have stepped beyond the little wooden gate by the coach house and walked the familiar pathway to the Honywood Oak and known exactly that thrill which comes with the opening of the gate, an anticipation of the sweeping peace that will fill me as I reach the oak.

'So you can measure, physiologically, the decrease in stress?' I ask.

He can. Mike tells me of a study conducted by a team at Edinburgh University where they had put mobile EEG sensors on various participants who were then made to walk about the city. As they went from urban to parkland environments their brains shifted from being in more stressed states to more meditative states. Hardly surprising, perhaps, but rather than merely intuitively knowing such notions, here was scientific backing. There was a neurological shift observable and people reported feelings that mirrored their brain patterns.

I am intrigued by that term 'meditative states'.

Mike explains. 'They're the neurological patterns that you'd see during a meditation practice. The amygdala calms down. The chatter becomes less. When you're in that oak tree, you're in an environment where you feel connected. Your brain finds it easier to attend to the stimuli around, so there's a subconscious easing which goes along with the "meditative state" notion. You're immersed and engaged with the environment – you might be looking at the birds, hearing their songs – so there is attentive activity but it's what we'd call "effortless attention" so you don't need to expend mental effort doing it.'

Being in the natural world, whether it was being by an oak tree or walking in the park, is really about the absence of the bad stuff, reducing the negative stressors that we have built, created – be they physical or social – in our lives.

'When we go to the oak tree, we escape from the negative stuff and return to our default, which isn't to be stressed. Most animals in their natural habitat aren't stressed. When we go to the oak, we're dropping the very recent addition to human life – the industrialised, nine-to-five existence that we've created.'

All of these factors lead to what Mike calls 'place-making' – we devise green spaces, which we then return to again and again. We see these spaces as safe, healthy places to which we can escape. In these places we gain the mental distance needed to allow us to reflect on our lives. As the artist Stephen Taylor did, returning endlessly to paint the same cornfield and the same oak. As I did with the Honywood Oak. And as I did later with the Field Oak and with the Stag-Headed Oak.

'So if you ask people when they're stressed, "Where would you like to go?" often they would choose to go to a nature space. They might say, "I want to go to the woods," or wherever. Because of that mental distance, the chance to reflect and the behavioural changes that take place when you're there.'

It certainly all seems like more good common sense.

'Thanks, Mike,' I say.

But I wonder if there might not also be other factors at play.

'So we've got these environmental, behavioural and neurological aspects. Are there physiological? Like you get increased oxygen in the woods . . .'

Mike nods.

'Sure. You might have heard of shinrin-yoku, forest bathing – specifically spending time in forests to increase your wellbeing? Well, there are certain phytoncides – chemicals given off by plants, and I'm sure oak trees give them off as well – which can affect our physiology. It's the basis for aromatherapy. Some of these phytoncides relax, some stimulate. Some even boost our immune system.'

Mike tells me of a series of studies that have been conducted in Japan where forest bathing was found to lead to significantly heightened natural killer cell function, which is a vital part of the immune system. Then the research team had replicated the study by pumping phytoncide into hotel rooms where they then got the same measures in natural killer cell activity. It was the phytoncide given off by the trees. There was a physical effect, a physiological effect, caused by the chemicals.[14]

'It is a fairly straightforward idea,' suggests Mike. 'Fundamentally, we have it hardwired into us – we know intuitively that it's good for us being in green spaces, being in the woods. That's why we're drawn there. For your oak tree, you've got the immediate impact of being in that place in an immersive, attentive activity – that doesn't mean you've got to be doing lots, the activity can be merely "I'm sitting here, taking in the fact that I'm here". You've got place-making and you've got the invitations to be behaving in calm, peaceful ways and the reinforcement in our brains that being there works: "Yes, that's going to be good for me."'

He relaxes a moment as though he is in a forest, bathing, rather than in a straight-lined, rectangular office.

'That's the glory and the challenge of the research that I do. Most would see it as common sense. I have to show it by the science. It's a straightforward idea. Now the science backs it up.'

28 June
Two Oak Hill.
I am seated comfortably within the boughs of the oak. I turn to wondering on those people that lived here, walked and wandered this land, thousands of years before. For many years, I have mused upon the ancient nature of this landscape. There is a wide, flat area that tops the hill. Surely people would have chosen to live up here in times past. There is the stream below and you can see the lands all around.

I look out from on high. The wind rises. The oak leaves bump and brush against me. The bough beneath my hand steadily warms. Two pheasants call throatily and fly beneath me. I feel the everyday become ephemeral. A lace-winged fly lands on the skin of my hand and some time later a tiny spider catches in the hairs of my right leg. I feel the moments melt. And as they do, so, too, something of myself also shifts and becomes less sure, less certain in its substance. I sit and feel myself gradually drift as though into some catatonic state, here upon this oak branch, my hands resting on two boughs, my feet on two others. And I start to wonder what would happen if I stayed here, for an hour, for two, for three. If I just sit here and let the day unfurl and the sunlight fade, while the clouds grow and rain falls as it is predicted to for later in the day, and then in time as the sun falls, too, and night reigns. And then as another day dawns.

And something within me desperately wants to answer to that calling.

3 July

It is not until late that I manage to rouse myself and head out. For some weeks a strange kind of exhaustion has befallen me, which has made each day seem momentous; simple tasks have become mountainous.

I walk out and plunge down the inviting alleyway of a path cut through a field of barley. The young plants are already waist high and I hold my left hand beside me so that the green beards of the barley brush against my skin as I cross the field. When I stop and look out across the wild, green expanse of the barley, it feels I am peering out on some mesmeric sea where the slightest winds cause the surface of water to shimmer. I imagine walking through here in the high winds of a storm and being enthralled by the sensations brought on by the sublime wonder of these fields, the barley pitching and rolling.

A whitethroat scratches away in the hedgerow a hundred yards before me. I feel a certain dream-like state befall me as I walk quite alone through the barley field, no other soul for miles about. The regular crump of my footfall lulls me as though into some narcoleptic trance.

4 July
In the early evening, I head down to the church with the same strange sense of otherworldliness upon me as yesterday. I walk the same avenue between the heads of barleycorn. The winds are fresh. I brush my hands against the soft green threads of the barley hair and watch the waves that flow over the surface of that green ocean. It feels in some odd manner as I walk that path that as I do so I am in some way changing, transforming from what I was to what I will become, though I have no notion of what that might be, only that there is within my being some deep and formative shift. Again, as I did yesterday, I step down the gentle incline of the field by the church towards that doorway through the hedgerow which marks the bridge over the stream. And today, it seems as though those rolling waters of the barleycorn have somehow baptised me, made me become something I was not before – a man from whom creatures of this

earth no longer flee; a man of whom they are no longer fearful.

13 July

It is early evening when I next visit the Honywood Oak. Though there are still signs on the road, the country show has come and gone. I walk through the wooden gate into the estate to find myself alone. I walk across the bridge beside the lake and up that familiar path to the oak. The undergrowth in the pines to the north has been dramatically cleared. I sit on the bench. It has been a long day. For a time, I simply sit. I let myself be. I breathe. I observe the oak. I breathe. My head lightens of thoughts.

The sky is gradually clearing and patches of pale, powder-blue sky grow, though the sun has started to drop to the west. I take off my shoes and then my socks and feel the freedom of the air upon my feet. Sunshine breaks from beneath cloud. I walk barefoot over to the oak and the bare earth of that mulched circle within the railed enclosure is soft beneath me. My soles sink into the soil. I stand on that seat on the western side of the oak, on the exposed cambium, smoothed by time.

Within a matter of a few moments being beside the oak, my cares and worries have been forgotten. I am living back in the present, as children do, as we all should seek to. My feet are planted in the soils beneath the oak boughs. My head is firmly back in the immediacy of the world about me. The truncated bough that points north a few feet from my head has a smattering of grey fluffy down that has been torn from the body of a fledgling bird. I think at once of the sparrowhawk, then of that other pale raptor which once soared over. It seems the bough has been adopted as a plucking post.

I sit back upon the bench and stay there a while. I think back some five years to the first time I came here, the first time I met Jonathan Jukes and talked of the idea of spending time with the Honywood Oak.

'I want to come and observe the oak,' I had said. 'Day and night. Sun and rain.' I would come whenever I could, in the spaces between work and home and spend the time watching the ways of the oak and learning of its life.

'OK,' Jonathan had said.

He had agreed to give me the freedom to come and go as I needed. I could park up at the estate offices then pass time whenever I wanted to beside the oak. A simple agreement. Yet one that allowed me the ease of access that I desired. Later, when I had driven away, tears came to my eyes in an unexpected wave of emotion that surged up through me. I had felt suffocated by the constrictions of my work and my home. Now, here was the opportunity to find the space where I might breathe freely. My hands had gripped the steering wheel as my chest heaved and my breathing wavered. I remember how the suddenness of the sensation had shocked me. I had pulled the car over and cried for the hope and possibility of the future. Then I had driven home.

Now I smile at the remembrance. When I rise, it is with the inclination to walk and so I head north around the edge of the lake, disturbing the greylag geese collected there in two great gatherings. They grumble at the effort to rouse themselves, but waddle away and drop with a series of loud splashes into the still waters of the lake. A grey heron flies, too, with rather more grace and elegance. I watch it as it flies north, following the reach of the lakes, and alights on a branch of an alder beyond the water and stands there in solitude, a sentinel figure.

Martins twitter above me. I follow the paths of the estate until I find myself back beside the Screaming Oak where a blackbird sings from one of the stag-horned branches that top the tree. More martins fly about above – martlets, guests of summer. I wander away west and find myself beside a still pond and look across the surface of the water to see a hare that stands on the opposite bank and stares back at me. In any such moments of connection with a creature of the natural world there is a sense of time stalling, of the second hand being halted. Yet there is something even more remarkable about a moment shared with a young hare – mythical, magical creatures as they are. We stay a while, transfixed by the presence of each other, then the leveret lopes gently south, the dark tips of its ears bobbing against the pale meadow grasses until it is gone and time ticks on again.

19 August

My head is heavy this morning, full of thoughts that swarm like bees within me. Unease creeps like a fever, aches in my neck and shoulders. Worry beats in my head. I know I must head out.

It has been some weeks since I have walked over to Two Oak Hill. Something in the shades of summer has shifted. The meadow field has turned a straw pale, the long grasses are now dried and dead and waving in the wind. The prongs of the stag head above the oak seem more prominent. At the top of the hill, the sandy slope is speckled with the gold light of ragwort. I walk to the Stag-Headed Oak and climb the now familiar steps up the ladder of boughs to my seat. My back rests against the oak. I sigh. I rub the back of my head on the bark, feeling my hair on the oak. I press the back of my skull into the wood and feel an easing. I can breathe. I will stay here.

Cries of voices by the stream fly up the hill. I startle. It seems as though I have woken to the sound but it cannot have been. I clamber hurriedly back down the oak to earth for fear of being found in the tree. Yet when I walk down the worn path in the sandy hillside there is no one there. I cross the bridge and drop a ritual offering of gold into the stream.

Keep my family in good health, happiness and hope, I say. Then I follow the track north, along the bank of the stream through nettles and willows.

It is my friend Dave Charleston who introduces me to Dylan Pym. I am chatting to Dave in his bookshop about my time sat beside the Honywood Oak.

'You want to meet Dylan,' Dave suggests. 'He loves his oaks.'

When I do finally meet Dylan Pym it is some months later. I head over to Dave's house, tucked by the green in the Suffolk village of Polstead, and he walks me the couple of streets round to Dylan's.

Dylan leads us through his house to a newly constructed room where he is busy adding the final touches to the woodwork. The plaster still feels damp. Dylan has been working on the skirting boards. In his hand, he holds a section of oak panel that has been steam-bent to run with the unusual curve of the wall. He bends down.

'Doesn't quite fit,' he says, looking up from the floor.

That is alright. We are heading down to the workshop anyway. I am due to be given a tour. Dave says he'll leave us to it. He's got two pheasants to pluck for supper. We say our goodbyes. I'll head back to his later.

It is a short walk away through the back lanes and fields of the village down to the sheds where Dylan works. To call him a wood worker wouldn't quite be fair. He is more of a wood magician. He steam-bends wood and creates wildly delightful and utterly unique pieces of furniture that have a fluidity to them unlike anything else I've seen. His cabinets, chests of drawers and bookcases have a surreal beauty. They have the solidity of wood fused with a liquid quality that bemuses the eye.

'I love oaks,' Dylan declares as we slip through the dusk light. 'There's the most amazing oak near here that I've been going to for most of my life.'

Here, clearly, is a kindred spirit.

Dylan was brought up in Polstead and he knows the lands about from nearly half a century of wandering the local lanes and ways and fields. Though born to academic parents, Dylan has travelled another path. Severely dyslexic, he has never learnt how to read or write. He still can't.

'Though I've got into audiobooks recently,' he says with a smile.

Yet he is without doubt one of the wisest, most knowledgeable people I have met on the matter of trees. And he is a genius with wood. He knows all the trees round about and is absolutely passionate about them.

'You've got to know them individually,' he says.

He is talking now of working with a tree's wood, of the importance of understanding the particular conditions each tree has known: the soil type it has grown on, its water supply, its growth structure.

'You can tell from the wood whether it's had enough water, what kind of life it's had,' he says.

We have emerged through an unkempt footpath into a grassy field where a line of trees grow right in the middle of the patch of land. They are a mix of birch and black walnut, planted there twenty years ago by Dylan.

'See how they've turned towards the water,' he says with a wave of his hand across the darkening sky. There is an underground stream that runs beneath the field to a far corner. Even in this light it is clear how the trees have gradually grown to lean towards the water. The effect is to produce a small copse that could have been a David Nash sculpture.

Dylan laughs at the idea he's produced some landscape artwork. He speaks with great enthusiasm for all manner of things. His energy is obvious. In the summer, he holds an annual festival on these fields where artists bring their wares and Dylan can show his latest furniture.

'Been having them for twenty years or so,' he says.

In his workshop, he explains more about how he works with wood – especially with oak – and how you can tell the nature of the wood with your hands, in the texture and touch and in the patterns of the tree's growth rings.

Dylan stands before me in his dirty, worn working clothes. His dark hair is rather dishevelled, flickered with silver like his unshaven beard. He rubs his hands together horizontally, as though one is a piece of wood. He speaks with the intensity of the fanatic.

'You have to know the wood to work with it.'

There is something deeply physical to understanding the wood. Seeing and feeling the complexity, the intricacy of the wood, the variances in growth through the years, is essential to Dylan's art.

Outside, on the ground beyond the workshop lies a series of sections of individual oak trees that have been sliced and stacked. In the coming months, some of these carefully chosen cuts of wood will be transformed in Dylan's hands to some remarkable new piece of furniture.

The light is gone and I am due to meet Dave. I follow Dylan back over the fields to the village as drizzle falls. He speaks of wood and oaks and I am happy to listen.

23 August

I rise early and long to step into the oak. So I clear my work and walk the road to Two Oak Hill. The sun is already hot, buzzards high and mewing. The moon hangs half-full in a cloud of pale, brushed wisps. Before the meadow I stop to watch two buzzards, which criss-cross every few seconds as they rise together before me. An audience of martins and swallows flies below them. The stag horn of the oak seems to have grown larger, though I know this cannot be. Yet more of the dead wood can certainly be seen, as though the tree is rising up from the sandy soils, reaching from its roots into the wonder of the firmament beyond. The oak looks magnificent — a wild beast standing proud upon the brow of the hill, looking out to the north, gazing over the meadows and fields and trees. I feel the shudder of a snake in the long grass.

I step up the hill and then on, up the bough ladder to my familiar seat within the Stag-Headed Oak. Yet no sooner have I settled there in that sylvan seat than I hear human voices from the east and something in them spooks me as before and so I start to climb down, afraid somewhere deep inside of what they may think or say or do on finding a man there within an oak tree. I clamber back to earth. Yet once there the voices seem to have vanished and I am half a mind to head back up. I stare into the oak and the thought comes to me that if the same is to happen the next time that I shall instead not clamber down but climb higher into the oak and so hide from human voices and eyes up there within the leaves and branches of the tree.

26 August

It is a golden summer's evening. On the other side of the road to Two Oak Hill, the fields lie shorn of barley, the stubs glittering.

The church bells ring out, lumbering through the thick air over the shallow valley. A woodpecker calls, which makes me turn my head back to the oaks. The woodpecker flies, diving south through the warm air, plunging into the dark green woodland that follows the stream. I study the trunk of the Stag-Headed Oak from where the bird has flown yet can see no nest hole. Though I want to climb up into the oak a while, I feel the presence of others soon to appear, dog-walkers treading the footpath that runs here beside the oak, and my hiding place, my seat within the Stag-Headed Oak is betrayed unless I were able to venture higher, to the top of the crown where the leaf cover is thicker. I know I cannot get there. The limbs are too small, too close together. Only a squirrel or a fairy or a dryad could do so.

21 September

An age has passed. I head into the oak and halt there, gazing out into the still, warm lands about. The corvids cease their cawing. A silence falls back upon the earth. The sparrowhawk appears, low in the sky, unfolding gently, gliding over the valley, turning wide circles before it falls by the poplars and is gone, out of sight. I see it fall and then hear the cries of buzzards to the east. I stretch my arms out along the oak boughs and close my eyes and know calm and peace. I feel the breeze and hear it rustle in the leaves about me. I am made lighter by being here. It is as though I have cast away some weight as I climbed. To be here is to be blessed. All earthly concerns remain below. Within the oak, the worries of the world below seem to leave me. Beyond is the distant clitter clatter of horses' hooves. Closer, there is the soft, contented cooing of pigeons.

28 September

The aroma of honeycomb greets me once more at the little wooden gate.

It is late afternoon. The sun of the morning has fallen behind grey cloud. It seems I am utterly alone in the estate. All others have left. My feet brush the gravel. I hear the water dripping down the stone steps by the bridge and the wind in the leaves of various maples that run down to the lake. I walk west, over the bridge and on up the incline to the Honywood Oak. I sense a stillness, an ease that befalls me and with that ease, sense, too, the tiredness I have been denying. At first I sit on the bench and for a few moments merely gaze back to the east and the blue sky on the horizon there beneath the grey cloud. For an hour or so, I simply sit upon the bench a discrete distance from the oak. Every few minutes, a smattering of seagulls appear in the sky above me; then they float away east, drifting with the cloud and the wind to who knows where. When I rise and head back to stand beside the oak, a greater calm has befallen me.

I walk with reverent steps about the oak, observing as ever the various aspects of the tree. I hope to find a fragment of fallen honeycomb but my search of the southern section of ground beneath the oak is in vain. After a single circumambulation of the tree, I walk back and over the bridge as though in some altered state of being.

I wander until the day has almost gone. Then I turn, back from the pine woods where I have ended up, and retrace my footsteps to the Honywood Oak where I stand in the gathering gloom, the gloaming, the fraying edge of day. I step closer until I am held in the darker light beneath the tree. I look up into the body of the oak above me and peer into the still shadows between the boughs and the leaves and the sky and for some time I stare up into that place and then turn and walk away.

A few days later, on my way to work, I get a text message from my friend Mark Mansfield, a lime plasterer by trade, a fine fount for musing on all matters archaeological and ancient. He tells me he has just bumped into the Anglo-Saxon historian Dr Sam Newton outside Mark's front door in the Suffolk village of Kersey. They chatted about local landscapes and then on the strange phenomenon of the grundles of Suffolk, odd gorges carved through the ground by glacial melt-waters, which Mark and I had spent a fine day investigating a year or so before. To some, these gorges in the ground may also be associated with the lairs of monsters – there is an etymological echo from the term grundle to Grendel, the arch-fiend from the Anglo-Saxon epic poem *Beowulf,* that is too enticing for some. Mark's message reads:

> We touched on the dangers of opening the witchy-poo doors to the witchypoo corridors.

I know what Mark means. Or, at least, I think I do. It is the temptation of turning from sensible thought to explain matters that are hard to comprehend. To understand those post-glacial features of the North Suffolk landscape as the lairs of monsters is certainly making a leap of faith. Modern geomorphology would argue differently. Go back a few hundred years though and the 'lair of monsters' argument to explain these strange, dark, watery places would have fitted rather well in many people's minds.

The desire to explain phenomena that are odd and unusual means that some, sometimes many, are prepared to turn to fields far from scientifically secure in order to gain some kind of explanation. It may feel there is no choice. Stepping from evidence of a rational nature to explain feelings and beliefs is, after all, at the heart of all religious thinking.

So how to explain the feeling of peace and calm that comes by being with, or in, an oak tree without stepping into irrational explanations? I think of the psychological explanations offered by Mike Rogerson, the reasoning of 'meditative states' and 'effortless attention'. They make good sense. Yet it feels there is something more.

For so long now I have been entranced by oaks, becoming oddly connected with individual trees and finding in their presence some kind of existential ease. However, I have no way of explaining this. I have no strong conventional religious belief system to turn to. So I am left to wonder.

Recently, I have been guided to the work of Monica Gagliano. Her research is dramatically shifting scientific views as to the nature of plant intelligence. When I reach my office, I remember her name. I open her website and the first thing I notice is an epigram:

Not all those who wander are lost.[15]

I smile. It is oddly reassuring. There is no reference to the author but I know they are the words of J.R.R. Tolkien, taken from a poem in *The Lord of the Rings*. On my desk at home, the same words lie scribbled in blue pencil on a torn piece of paper, penned a good few years ago by my then five-year-old daughter Molly. She had morphed the quote to 'Not all those who wonder are lost'.

The work of Gagliano is startling. One of her papers has the title 'The Mind of Plants: Thinking the Unthinkable'.[16] The essential point she makes is that plants have been shown scientifically to be capable of learning by association. They can choose.

Nor is Gagliano alone. There is a separate collective of scientists calling themselves plant neurobiologists who are also starting to probe into notions of plant cognition and sentience. I have just begun to read 'The Philosophy of Plant Neurobiology: A Manifesto' by Paco Calvo who is working with a team at the University of Murcia in Spain dedicated to studying the ecological and philosophical basis of plant intelligence. The opening line states: 'Speaking about plant intelligence is not taboo any longer.'[17] This strikes me as an obvious link to the pioneering work of ecologist Suzanne Simard back in the 1990s in British Columbia. She established the concept of a mycelial web, a 'wood-wide web', which exists in the wider reaches of trees' root systems, operating as a connective network and a vital life-force of the forest. It was Simard's research that showed us that trees are not lone individuals. In her words, they 'communicate with each other above and below ground'.[18]

Now we are beginning to move towards understanding something of how plants and trees may think.

It is all rather wonderfully mindblowing.

I step out for fresh air and by a serendipitous turn, I bump into my friend Sarah Beavins on some stone steps. From various conversations before, I know Sarah has a deeply developed appreciation of the sacred aspects of the natural world.

'I'm just off to get a coffee,' she says. 'Can I get you one?'

I've known Sarah for a few months. She is a bundle of vibrant, infectious energy. I gladly accept her offer and we arrange to meet in a few minutes.

When she arrives at my room, it is in a busy whirlwind of bags and coffees and smiles. She is planning to write a series of pieces forged around the elements.

'One would be on windfarms,' she declares. 'And another on sewage works.'

We laugh and joke together on how to write on matters of waste management.

She knows of my time getting to know oaks. I'd asked Sarah some time back if she would guide me into her spiritual beliefs. Now I am keen to know more on her thoughts around what it was that I had really been practising all those times with or within oaks. I start to broach the subject. Was it a form of tree worship? It didn't feel like oak worship was what I had been seeking to do. But now, looking back, it certainly did have something to do with some kind of meditative practice centred on being in the presence of oaks. But had it really been about becoming the oak or merely being in the company of oaks? I pause and frown. It is hard to find the words.

'In meditation, what I would seek to do is to become a birch tree or an oak,' she explains. 'Sinking my fingers into the bark, keying my fingers into the lines of the bark, putting my forehead against the oak and breathing . . . I do seek to become the oak.'

'Right,' I murmur.

Sarah continues, 'You know what an oak is going to feel like when the tree puts forth that fresh green at the beginning of May, that delicate light green. For me, I certainly try to become the oak or the hawthorn and I try to leave myself behind. I think we are all part of the same whole.'

Sarah smiles.

I remember Monica Gagliano's line about 'thinking the unthinkable'.

'We are part of the same whole,' Sarah repeats.

'As living creatures?' I venture.

'As living creatures. There are only so many molecules in the universe. It's a closed system.'

'Would you call what you do specifically a pagan practice?' I ask.

'No,' says Sarah.

'It's a meditative practice,' I begin. 'But it's not really Buddhist either, is it?'

'No. There's no particular strand of paganism that says "do this" or "do that".'

It is fascinating. Terms such as pagan and paganism have attracted negative, pejorative notions in many people's eyes. But I remember the etymological truth that the word pagan merely comes from the Latin term *pagus* meaning 'of the countryside'. If you live in a rural world, you're pagan – by definition. Surely, we would all happily accept a sense of being connected to the natural world.

Sarah jokes that she described me to a friend in an email some days before as 'rooted in landscape' and 'a bit pagan round the edges'. I like that.

I start to try to frame what Sarah is saying with respect to what it is that I have been doing, from those days passed walking around the Honywood Oak, seeing the tree, learning of the ecosystem of the tree, knowing the individual oak tree. That had been one part, it felt. It was a process of learning, of writing about the creatures that made up the ecological community of the Honywood Oak. In that time, I had discovered the calming effect that being beside an oak tree has. I had been thrilled and delighted to step through that little wooden doorway into the estate grounds, knowing that I was simply going to sit and observe beside the ancient oak.

Then there was that other layer to my experiences with oaks – the more deliberate act of sitting, spending time – meditating, if that's what it really was – by the Field Oak, and of climbing into the Stag-Headed Oak on Two Oak Hill.

'I wasn't trying to become the oak. I was still a separate being,' I say.

Sarah has some advice.

'Drop the intellectualisation. Just seek to become.'

I smile. I'm sure it is good advice, too.

'Trying to be without thought is part of what the meditation is all about,' she explains. 'To leave behind preconceptions. Not trying to be saying,' she put on a rather more playful, whimsical tone to her voice, ' "I wonder what it's like to be like an oak. I wonder what that oak might say to me". But instead simply saying "I am".'

'I guess what I'm trying to get at is why it was that I was made calmer and happier and more peaceful in the company of oaks,' I say. I am thinking of the words of Mike Rogerson. 'It's not an easy question to ask or to expect an answer from,' I say. 'That's why I'm sending out my feelers to various people who might answer it in different ways.'

Sarah sighs, nods gently, and gives a sympathetic half-smile of appreciation.

'You're next to a being that will live for hundreds of years. It gives us a sense of our own impermanence.'

She points to the ceiling.

'Like when you look up there, at night, and realise we really are so tiny.'

I nod. I know what Sarah is getting at.

Here, no doubt, is as good an answer as any I will ever get. To gaze at the stars, to be in the company of oaks that will live so far beyond us, gives us peace, gives us the capacity to see ourselves in the context of a world so much more infinite, more significant than our own.

Later that day, I turn back to the books.

I have had Stuart Piggott's *The Druids* out of the library for over a year. Piggott talks with the voice of the rational academic as he outlines how the problem of 'our knowledge of the Druids' sits alongside the problem of our 'creation of Druid idealizations or myths'. Our knowledge of Druids is from patchy classical texts and our vision of them is sometimes overblown and colourful. From the evidence, the Druids can certainly be seen to have existed as a 'prehistoric Celtic priesthood'. But there is a distinct division between what Piggott calls the 'Druids-as-known' and the 'Druids-as-wished-for'.[19] Any truths they might have possessed, including their worship practice centred on oak trees, has gone. Our present vision of the Druids is one re-constructed from their Romantic rediscovery in the eighteenth century, seeing Druids as sages, as ancient Britons with sacred knowledge of the ways of the natural world that we have lost over the last 2,000 years.

I close the book, shutting the words within the worn green hardback covers. I push it across the desk and sit back in my chair. I think of Sarah's comment from earlier.

'Drop the intellectualisation. Just seek to become.'

I am seeking oak knowledge. I have experienced what so many other people know – that being by an oak tree has some kind of calming effect. I know this. Whether

I am staring at the ancient glory of the Honywood Oak, or sitting by the far younger frame of the Field Oak, or tucked in the boughs of the Stag-Headed Oak upon Two Oak Hill, being with those oaks has made me feel better. My mind is stilled. I feel less anxious. I feel more secure in my own skin. I am able to accept the present, am able to be myself more easily. And in those moments I do not seek answers as to why that is. I simply accept the composed balm that the presence of oaks offers.

4 October
I stand upon Two Oak Hill as a brisk breeze blows in from the east. The day is glorious. Sunshine and pale, pillowy clouds fill the sky. I let my feet lead me until I am up once more amongst the boughs and the leaves, lifted from the land and held there in the Stag-Headed Oak. Strange how the oak boughs seem like human limbs.

A certain something held me earthbound for a few moments before I stepped beyond and climbed into the oak. And once here, inside this realm, a feeling of calm soon fills me. It is a sensation only tempered by the thought that I cannot be seen here, cannot be caught by someone passing by who might find me. That concern remains with me as I sit in the oak and I cannot shake it clear, cannot simply be there in the tree unaffected by the fear of what others may say, or think, or do. A grown adult sitting in an oak on a hill.

The wind blows. The acorns shake and the smaller boughs shift yet I am secure and firmly held and feel my back warming against the trunk. I grow calmer. My worries are gradually forgotten. I no longer feel alone.

Within the oak, within this world up here, the world below is far away.

11 October

I head for Two Oak Hill. Though I only have an hour or so, it feels the best thing I can do. I feel I have to go there. For a matter of moments I will sit within the oak and all will be well. And so I do, and there, within the tree, I feel peace descend upon me. The pressing of the boughs against my back and my head are a sweet relief.

I know not what to do any more. I am earthbound and seek to fly. Each day, I seek to rise, seek to be within the oak – to hide there and leave this world behind. Yet it cannot be. And to do so is simply to flee. So instead I clamber, when I can, up and away from these lands into that seated sanctuary in the Stag-Headed Oak upon the hill. And there for a moment there is peace.

18 October

On Two Oak Hill, the wind holds a chill. I walk west on the path past the grey statues of dead thistles that shake in the cold wind. On the peak of the hill, I brace and rise to my full height and gaze out across the treescape that in its fringes is turning to browns and yellows, to hues that tell of the season to come.

The two oaks are touched, too, with autumn. The oak leaves seem to shiver in the face of the wind. They have shrunk in size as they prepare to lose that green of life and gradually die. When the sun surfaces from behind one of the last of the grey rainclouds, it still holds enough warmth to remind me of summer. The rays give the softest of touches then flee again beneath cloud.

By the furthest oak, two owl pellets lie sodden and blackened by the rain. I gaze up to the scratched surface of a branch twelve feet or so from the ground that seems to be the site where the owl spends its nightly vigil. I will come here, I declare to myself. Soon, I will come and sneak my way through darkness along the path, or at the gloaming time, to catch a glimpse of that white-winged angel of the night.

I sense that time is pressing. Yet as I walk back past the Stag-Headed Oak, I cannot resist and so step up into the tree to sit and watch the sun's rays reflecting on the oak leaves. I stay for a few fleeting but blissful moments and rest there and wonder what it might be like to have the time to stay and watch the leaves turn from green until they fall, and remain upon this bough or others hereabouts until I become as integral a part of the oak as the limbs that hold me here.

I am due to be joining my good friend Paul Gwynne on a journey out to the lands beyond Gosfield, a couple of villages west of my home, to visit Stephen Westover. We are to pick up some 'hethering' for the gardens Paul manages. He used the word in an email the other day and I don't know exactly what it means. But I do know that I want to meet Stephen. Dylan Pym declared that if I wanted to meet someone who 'knew more about trees than most others around here', then I should seek out Stephen Westover. Stephen runs an arboriculture and sylviculture business with his daughter Becky. Together, they manage several small woodlands, working the wood they cut down into anything from firewood to cleft oak roofing shingles. They are based in a field near the village of Blackmore End.

'Wrap up warm,' Paul warns in a text. 'Their base is an open-sided barn and very exposed to the wind.'

Lunch at Paul's is a homemade beetroot-and-squash soup that is the same glorious, deep, blood-purple colour as sunrise had been this morning. We settle down at the table, grateful for the hot broth.

'So where are you at with the oaks, then?' Paul asks.

He has followed my progress over the years from my first steps towards the Honywood Oak to the more ethereal ground I now walk. I try to explain how I am still trying to explore the relations between humans and oaks.

'Well,' I venture. 'Looking into what I guess you could call the spiritual side of oaks. And how people who work with wood feel about trees.'

Paul nods.

'Something along the lines of why it is that many of the people who work with wood – even some of the toughest woodmen – seem to have genuinely emotional feelings towards certain trees.'

I tell the tale of the woodman Richard Fordham speaking fondly to me of a particular favourite, an individual willow tree. Paul knows Richard, too. He raises an eyebrow over his soup bowl.

Paul is good to talk to about such matters. He has spent many years working the gardens and the woods on the small estate which he manages, so he is properly grounded. But he is also a philosopher, a surfer, a man who appreciates connecting with the natural world. He understands the difficulties of seeking such answers, of opening such devout doorways. And he also knows the need to ask the questions.

I mention the words of advice from Sarah Beavins a couple of weeks before – not merely to think about oaks but to seek to become the oak. They have kept ringing in my head. Isn't that what I have been doing all the time that I have gone and sat for hours by the Field Oak, or as I keep climbing into the Stag-Headed Oak on Two Oak Hill – if not becoming the oak, I've been actively seeking to be somehow transformed by being in the company of oaks?

'Well, the scientists are now telling us all about the ways that trees communicate with each other on the mycorrhizal level,' Paul says. 'And there are others who have taken acid in the woods, who will tell you that trees are alive in ways we don't normally appreciate.'

We laugh.

'They're not so far away from each other,' Paul adds.

We laugh again. It is true. There are many paths to seeking the truths about oaks.

It is an experiential thing. Spend time with oaks and you will start to appreciate aspects of the trees that much of our modern culture has forgotten but which many woodmen and woodwomen, and wood worshippers, will still feel. And some of which both modern science and nature-based theologies will also agree with. It then comes down to how you want to express the matter – with words that tell of notions of neurological and behavioural patterns or with other words that tell of becoming the oak.

We head off in the Land Rover and find Stephen and Becky down a track in the flatlands of North Essex, beside their open-barn workshop which is stacked with walls of firewood. They are both layered up against the wind.

Paul makes some introductions. We joke about the cold.

'I've got seven layers on,' Stephen says.

While Becky and Paul chat and Becky's two bounding sheepdogs – Splinter and Chip – make their re-acquaintance with Paul's working cocker spaniel Berkeley, Stephen takes me for a walking tour of the workshop. He explains the various traditional woodworking techniques they still use – such as cleaving the wood by splitting along the grain rather than sawing, which means keeping the strength of the wood. They use tools such as mauls and froes. He points to a pile of split laths which would be used to frame the walls of timber houses. Becky has made them.

'She's much better than I am,' he happily confesses, with a smile of pride.

It really isn't that usual to find a working woodwoman. I look over to Becky who is enjoying her lunch break, eating sandwiches while leaning on the back of her truck. Their business is safe in her hands – as are the local woods.

Stephen has a sense of clear-eyed wisdom. He speaks with a calm, level, unrushed manner that tells of years of experience and knowledge. At one point, he calls himself a 'scruffy woodman'. He may be that but he is also someone who knows his trees. He knows the Honywood Oak very well. He was working on the arboretum at the Marks Hall Estate over thirty years ago, even before Jonathan Jukes arrived there to oversee curation of the trees on the estate.

'We built that protective railing around the Honywood Oak,' he says. 'About four years ago.'

Now he mentions it, I can see the delicate care that has clearly been taken to build the low barrier that encircles the oak, politely preventing people from standing too close to the oak and unwittingly compacting down the earth around the tree. I can picture the sections of worked wood that form the horizontal hurdles which are like longer laths, the grain of the wood running along, each section sculpted, not severed, from larger lengths of wood.

We return to Paul and Becky and together venture over to a black, corrugated farm silo. It turns out to be another wood store – containing slices of sweet chestnut which lie stacked in three-metre sections. Paul is planning to build a jetty on the edge of the small lake back on the estate at Alphamstone. He steps into the silo to check out the wood with Stephen. I get the chance to chat briefly with Becky. She is wrapped in a worn blue jacket over layers of working clothes. She's been working with her dad for some years now. They make a good team. You can tell how well from the easy manner they have together.

'Time passes,' she says wistfully.

We both peer into the darker corners of the silo.

'You can smell the rats,' she says.

It is time to go. Becky has to collect children from school. Paul and I both have to be back to pick up our daughters. But first there are five bundles of hethering that need to be lifted onto the roof of the Land Rover – nine- or ten-foot-long lengths of hazel, which can be used for weaving together young hedges to support their structure. Paul will use them to create garden features – flowers can be grown up through the mesh of branch and twig.

In the Land Rover on the way back, Paul and I talk about how grounded Stephen and Becky seem. They are earthed. It is as though the time spent in the woods, with the trees, has done something elemental to them, tethered them. Each has such a tie to wood. They cut wood, work wood and have become somewhat of wood – they are wood folk, connected to the earth.

I look out the side window over arable fields, picking out the patches of woodland against the grey skyline. Even only a few hundred years ago, before industrial times, so many more of us would have lived rural rather than urban lives. We would all have been much more involved with woods, so much more used to having trees in our everyday existence.

25 October

The sun rises. The field is silent, sleeping on under a pale layer of mist that floats some four feet from the earth. As I walk the rabbit track through that diaphanous shroud, it vanishes about me, falling away into nothingness. Yet as I reach the wooden fence at the far end and turn, I see that same ethereal mist is still there. From the fence I can see the sun, now high enough to emblazon the upper branches of the trees down the road. Its mellifluous touch brushes the mist. The world awakes. Birds fly. Gulls, pigeons and crows appear in the hushed skies. Three horses graze silently.

I stay a while and watch as the sun grows and the pale vapour becomes ever more honeydewed as it warms and evaporates.

An hour or so later, I stand beside a hedge and listen to two of the last sparrows. When I walk away, it is towards the oaks. I stop beside the graveyard where blackbirds call and fly among the stones. The sky is layered with a sallow white sheet through which the sun will break, lighting the autumn leaves to golden. The chestnut trees have turned entirely to shades of dun browns. The conkers lie scattered on the verges.

There is no one about. A stillness has fallen on these low lands. I cross the bridge over the sacred stream and rise towards the two oaks. I give the only offering I have: an acorn that has been tucked away in a pocket for an age.

I see oak trees everywhere. On the side of the road as I walk the lanes of the village. I plot the oaks that I know into an oak map of the landscape. I see them on the side of the motorway as I drive back from London, shining gold oak leaves when all other trees are bare. And I see oak leaves everywhere. Opening the front door in the just-born light of a new day, scuttling my two daughters, Eva and Molly, out towards school, a clutch of oak leaves lie on the front path like some votive offering. There is often no time to stand and stare, only time to glance and register the presence of oaks still in my life, always there on the edge of consciousness, leaves glimpsed out of the corner of an eye. One day, an oak leaf – worn and muddied – sticks to the front bumper of the car. It is a sign for those who care to see. I see it. On another day, ascending the flights of concrete stairs in the library – a single oak leaf is there upon the bare floor, four levels up from the ground. I cannot help but pluck it from the stone floor and place it as a sacred bookmark in the book I clutch beneath my arm – Stuart Piggott's *The Druids*. I cannot leave the oaks. Either they will not let me or some part of me will not let go.

Later that day, I step through the gate into the Marks Hall Estate and am held once more by the distinct scent of honeycomb that seems to fill the air. A strange shrub whose name I do not know buzzes with the endeavours of bees as though it is spring. Bunches of tiny, pale, lantern flowers still hang, offering nectar. A strange crop of orange-pink fruits hang, too. I pluck one and walk over lawns littered with shrunken apples and pears.

I cross the bridge over the stream beside the lake.

I walk to the south of the Honywood Oak and over the low wooden railing until I am beside the tree. A fresh section of honeycomb lies on the floor. It is an offering. I lift the treasure and am thrilled at the sight, delighted in its delicacy, its fragility. I lift it to my nose. It smells smoky sweet.

Later, a heron flies north up the stream, flying with such light grace yet landing so leaden-legged as it falls back from air to earth. No birdsong breaks. It feels as though the world, these creatures hereabouts and I, all await something. A pheasant silently stalks the land before the oak. Two, then three of the geese arise and shake their feathers in white waves and settle back down. The heron rises and flies north again, up the line of the stream. I, too, head north for a while, to seek the sound of goldcrest amidst the pines.

In time I walk back to the copse of crab apples and shrunken pear trees and with a last glorious catch of the mysterious honeycomb musk, step through the small, wooden gate and am gone.

8 November

There is an icy touch to the cold in the north wind. I step out. Pale patches of cloud bank in the sky. A tractor runs along the field beside me. In time, sunlight breaks through and warmth comes. Six white doves fly above the surface of the field and startle me with their presence as though they have blown in upon the winds from some strange and distant land.

I walk south and scrunch my eyes to the sunlight in order to see. The day rustles and crumps. Dried leaves. A distant gun. The day clears. Pheasants croak. A mistle thrush flies high and wide about the church. I walk on south. Two crows sit upon the branches of the stag horn. As I step into the meadow, one flies, the other remains. Sunshine holds. The day clears. Upon the hill I stand awhile and watch a stream of black rooks lift from green fields far beyond the farm and run east through the blue sky above the woods.

I slip up into the oak, climbing easily and tucking down, facing south, sheltered by the trunk from the worst of the cold from the north. Sunlight splinters through oak leaves and falls upon my face. My head leans against the oak. My eyes close. For some moments, I remain there, huddled limb to limb with the oak, high above the ground, enveloped in the tree boughs, in the leaf sounds. I feel a gentle warmth and calm and peace steal upon me.

13 November

I stop beside the church at that familiar space beyond the hedgeline of the graveyard. The sunlight is warm. I stand and speak to my father. A mistle thrush flies in from the south, from down where the oaks lie. I see it from afar – a bird at pace, pale-chested, heading my way. And as it turns and rises above the hazel tree beside me the bird flies exactly the path that it or another mistle thrush flew a few days before when I stood just as I stand here now. And I know that the sighting has to be auspicious.

18 November

All over the land the oaks now burn with autumnal in-tensity. By the time that I reach the hill, the sky is wide open. On Two Oak Hill, I stop and stare to the south to where strange archipelagos of cloud islands have formed at the far horizon that could be distant continents on some lost sea. I walk towards the Stag-Headed Oak whose wooden antlers seem today to strike even higher into the grey sky above. I halt hesitantly a while then plunge into the brittle covering of the bronze leaf canopy and touch the oak and climb again into the tree's sweet relief. The ground below has a layering of leaf fall. Today, the wind blows up the hill from the south-east and I cannot hide from its cold touch, so I brace myself and face the frozen air until my nose grows numb.

25 November

From the ridge of Two Oak Hill, I can hear the mewing of buzzards to the south, somewhere beyond the line of poplars, beyond the streak of grey smoke pouring from a distant fire. The church bells ring twice. I stand on the top of the hill and see how the oak has thinned in the last week and is clothed in rags now, bare on top and exposed to the winds of winter.

I climb into the oak and shelter there beneath the boughs as the winds blow ever fiercer and leaves break free about me and I feel as though I am aboard a boat upon a sea in a storm, the scene lit with a glorious intensity by the low winter sun.

26 November

Late on in this short day, I come here to the Field Oak. Bronze leaves above me flare fiery, burning brilliantly in the last embers of the day. In time I settle and become a part of all around me. A robin arrives and flicks about the leaf litter. His bright bib matches, every few steps, one of the shades of a fallen oak leaf. A grey squirrel climbs the branches of the oak, agitated, nervous, with constant twitching of its tail, then stands squat and cries and shrieks. It is the strangest sound, a primal scream of some sort that unsettles my calm. For some moments, the scream pours forth and then as though alerted by some call from home, the squirrel stops and scampers off along the branches, skipping neatly, slipping into the hawthorn, and is gone.

I settle back into my seat and the silence. Small birds flicker and fly and the last of the sun's rays touch the topmost branches of the oak. I watch from beneath. As you stay and sit and sink into the essence of all that is of the world about you, so your own presence fades. A mouse appears as a rustle in the leaf litter and finds form as a grey shade in the dusk, then merely is a line of sound that runs upon the ground three feet from my feet.

Night does not fall but creeps in, seeps in, as the light fades. The darkness and the silence grow. Cloud drifts over, hiding the starlight. I sit beneath the oak and try to simply be. Soft winds ruffle brittle leaves. It is the only sound. The birds of day have all gone quiet. I hear no creatures. No owls call. Moments pass. Nothing happens. Something stirs within the oak. Something falls, clattering to earth. And as more time passes, deep silence settles. I look to the sky and the cloud has cleared and starlight shines anew – pale orbs of light beyond the dark canopy of the oak. A fox barks in a distant field. A breeze blows, shaking dry leaves. More moments pass. I hear the steps of creatures whose forms I cannot see. Another wind rises, disturbs, stirs another susurration of the oak leaves. I listen to those oracular voices. I close my eyes and feel the darkness close upon me.

30 November

By the time that I reach Two Oak Hill, the sunlight is turning old. A golden touch falls on the hillside. The remaining oak leaves are brittle and metallic. Most now lie upon the ground, a bronze carpet that shines where the sunlight falls. I step about the oak, then climb up into its sunlit heart and sit and feel how bare, how exposed the tree is now that only fragments of the leaf canopy remain.

Something has begun to change within me. Something in this seeking out of oaks has transformed me into someone better.

The cold creeps in quickly, sending me clambering back down. I seek home before the sun has gone.

I wind my way back north through dwindling light until I smell the oak smoke from the bonfire by the church and stand and wait there a while and marvel at the colour of the flame, how it is matched, mimicked by the colour of the setting sun. The day is dying. A mauve miasma fills the sky. As I head up the hill beside the church the scent of the sacred incense of the oak smoke grows ever stronger. I reach the oak beside the road and a hawk falls from the bare finger of an outstretched limb. Its bronze back glints in the sun as it flies south. I walk on and turn into my road. A deer crosses my path ahead and stands beside a shrunken oak and then turns and stares at me when I am only feet away and holds that stare unafraid before turning again and vanishing into the field.

11 December

I sit beside the Field Oak and feel the fervour of the fanatic. My soul seems ever more in communion with whatever spirit or being it is I contemplate before me.

Starlings stand among the higher branches, then shoot into a gunmetal sky. The fires in the west are dying down. The pheasants call the tune of dusk. Others join in. Black-birds seek to settle. I know the patterns so well now. It is the calling out before the quiet, before the light goes out.

14 December

The moon is a vast white orb tonight. A circle of light. I watch from the oak in the field. The moon shadow haunts the field and flickers in fragments on leaves of bramble below and on the remaining oak leaves around me.

Earlier, I watched the moon as it climbed the eastern horizon. Then it was even vaster. When it first rose, it was brushed with soft hues of honeydew. Now it is paler. I will watch it as it ascends the sky, as its ashen sheen grows ever whiter.

22 December

A rug of frozen oak leaves lines the floor of the field. The sun rises golden on this solstice day. It is the turn-ing time. Light gathers. Sunshine seeps back into the day. I know the world as never before. The oak is warm beneath me. I feel its skin on mine. I close my eyes and face the winter sun.

When I return to Dylan Pym, he is enjoying lunch in the warm, open kitchen of the thatched cottage where his mother and father live, just down from his workshops in the backwaters of the Suffolk village of Polstead. There is an idyllic quality to the cottage, tucked away down a lane and painted a deliciously warm orange colour. Dylan's son Dan is there, a pencil poking out of a woolly hat, his fingers working neatly away at rolling a cigarette. Along with a friend, Carl, he's been busy building the wooden frame for what will become another room at the back of Dylan's workshop. They've all just finished eating. Homemade squash soup still lies in a saucepan. It is a softer shade of the delicious carroty ginger of the cottage's outer wall.

'Would you like some?' Dylan's mother, Paule, kindly offers.

I say I am fine but happily have a tea instead and join them around the kitchen table. Dan tosses a rollie over to his granny. She is busy talking European politics with Carl. When Carl, Dan and Paule head outside to smoke, Dylan and I turn back to the oaks.

Dave Charleston – local bookseller extraordinaire –

has told me of an oak nearby that Dylan is known to be especially fond of.

'It's a twenty minute walk from here,' he explains. 'Not far away. But on a country estate so not easy to access. There's some ancient oaks alongside some younger ones – used to be a deer park. I found it because I used to be a shepherd. I had a hundred and fifty ewes and used to rent the land for grazing.' He pauses.

'I was a kid, well, in my twenties anyway. I started to climb all the trees. I found this oak – no one knew about it until I started to blabber on about it and people got to know. But this tree . . . it's just the closest I've ever been to being a tree.'

I look across the table. Dylan's pale blue eyes are sparkling as he tells the tale.

'So basically you've got this struggle to climb it. Not a massively big tree but totally rotten inside. What you've got to do is climb it and then you can drop right down inside it.'

He halts.

'Well, of course, first of all, you start freaking out, like, "How am I going to get out! How am I going to get out!"'

His voice mimics the panic.

I laugh. I have an image of Dylan standing in the dark, in the hollow body of his oak.

'Cause you're right down inside. You're immersed right inside the tree.' He shifts in his chair, looks away in reminiscence. 'Then all of a sudden, everything just goes . . . *pewwww*!'

He blows out softly. There is a glorious intensity to his face as he tries to get across the profound emotion of that moment. 'You get this beautiful feeling . . . like, "I'm inside this two-hundred-and-fifty-year-old oak," and then this whole energy just hits you. Bam! And you can just stay in there for ages.'

'Wow,' I say.

'You can see the hole in the top above you,' Dylan slows. 'It's not that high . . .' he looks to the ceiling of the kitchen. 'About as high as that beam . . . maybe twelve foot.'

I look up.

'Right.' I add, 'You're completely inside it.'

'Yeah, you're right inside it. And you just start feeling it . . . this energy.'

He pauses and leans back in the chair.

The back door opens and Dan appears with a stack of firewood for the range cooker clutched in his arms.

'So I started taking people to the oak,' Dylan continues. 'People who had problems, mental problems of some kind. People who were uneasy about life or something in their life. And a lot of people didn't even get to climb it. I mean, it wasn't easy. And then there's the psychological battle of dropping down into what's inside it, because once you've overcome that . . .'

Some people hadn't been able to take the leap of faith required to jump into the oak. Others had.

'The whole thing was about being on your own. You had to drop down inside it and be there on your own.'

It is such a strong image. There is something distinctly mythological to the manner of the transformation that Dylan describes. To step within the realm of the oak and then to emerge again washed of the ills of before.

Dan elbows his way through the back door with another armful of wood.

I think of a moment the day before when my next-door neighbour at work, Chris McCully – sage and wise knower of all manner of things – appeared at my office door.

'Adonis,' he stated with grave intent.

It was a follow-up to a chat about oaks and trees and mythology. And now I think of Adonis once more – born from a mother transformed into a tree.

Dylan fills the moment of stillness.

'It's just such an incredible feeling . . .'

He takes a pause to find the right words. 'Peacefulness . . . calmfulness. No anxiety. No angst.'

That is pretty much it, isn't it?

Peace. Calm.

Those are the two words that I find myself coming back to, time after time, to express the feelings that I experience when I step into the dominion of the Honywood Oak, or sit before the Field Oak, or more emphatically, as I settle upon that bough in the Stag-Headed Oak on Two Oak Hill. Peace and calm.

To hear this woodman before me state those exact two words to describe his feelings on being within that oak is fantastically affirming. And I know I can happily leave it there. We can rise from the kitchen table and head outside to join the others. We can step into the glorious Suffolk lands beyond the kitchen door. But Dylan is fired with a passion.

'I wouldn't really call it a spiritual thing so much as a connection to nature,' he says. 'That's what I get out of it. I think what you've got to do is get people to appreciate that a tree has such a powerful energy to it. That's the spiritual side to it, or whatever you want to call it. And then there's the other side to it – working with the wood and the beauty of wood and knowing what it can do and can't do. And the individuality of each tree – where it's grown affects how you can work with it later on. You can feel it. You instinctively feel it.'

His hands rub an imaginary piece of wood.

'Like you might feel some brittleness, feel a knot and subconsciously think, "This one's had a hard life," or whatever.'

Paule comes back into the kitchen. I can see Dan and Carl heading over towards the workshop. It is time to let Dylan get back to his woodwork. But he has enthu-siastically launched into another oak tale.

'Some of the other oaks down near the Hollow Oak are fantastic,' he declares. 'Some of them are at least seven hundred years old. I get everyone to hug a tree, but to do so by reaching round and touching fingers . . .' He explodes into a mixture of laughter and joyous reverie. 'I mean, of course, I know because I've done it so many times before, but I'm like, "Come on! How many can we get round it? Come on everybody! Touch fingers!"'

And I can see Dylan with his clutch of friends, each stretched with arms wide, splayed around the bark of an ancient oak tree, a circle of humans enwrapping the oak.

'That's such a great thing to do.'

'It isn't so much about hugging a tree as showing the size of it, to get people to appreciate how big this tree is. I mean, there's six of us, finger to finger.'

I do the maths.

'That's eight hundred years old,' I say with some authority. I know from measuring the Honywood Oak some years back, though not with human arm spans. 'Twenty-eight feet is about eight hundred years old.'

It is me who is now enthused.

'And as soon as you do something like that with an individual tree, you do build a connection.'

I picture the Honywood Oak surrounded by a handful of children, each hugging the tree, their fingertips just touching.

'Yeah,' Dylan agrees. 'For me it was about trying to get *other* people to see that, trying to get them to engage with that sense of connecting with the oak.' He bursts into laughter. 'I mean, I'm there anyway!'

Dylan's words fill the kitchen. His pale blue eyes laugh as he does. He certainly is there.

We walk over to the workshop – following the track across the stream that runs beside the cottage, through one wooden gate and across the long paddock containing two dozen sheep dotted about the field. There is something deliciously idyllic to the scene. We talk for a few minutes more about his plans for the future then I let him return to his work, to more pressing concerns – there are eighty Polstead chairs that need construction.

As I leave and walk back to my car, I can't help smiling gently to myself. Here is a man who understands the simple truth of what I am trying to get at and who really knows the feeling of being connected with an oak tree and the peace and calm that result.

ENVOI

S weep the night from your eyes. Step out into the cold darkness. The day is almost upon us.

Duck down behind the framed border of the yew hedge, the soft slopes of the valley. Walk over the bridge as the song of the stream sings through the lingering grey motes of night. The path is pale, lit by the gibbous moon. The shadow cast is clean and clear. Listen. All is still. All is quiet, hushed in the expectancy of night's end.

On the edge of the lake, tucked in silent prayer, stands the heron. Aged, hunched, ever in hermitic reverence, the solitary guardian of this sacred space, ghostly in his grey shroud. In this liminal light before dawn, his presence is otherworldly and haunting.

Light gathers. The oak stands massive, iron cast in this fading darkness – a vast, round-shouldered presence. A breeze stirs the leaves, a breath that seems to seep from the dawn. The barking of the deer destroys the silence with a fierce, devilish cry. Then it is gone and the vacuum is filled with peace. The blackbirds flit and twitter about the base of the oak. Their voices, like the wrens that chit their Morse-coded calls, are soft, stilted and reverent of the moment. I watch as a single white feather falls from above into the leaf litter. And the call of the deer shatters the silence once more.

The sun rises. All awakes. In the east, grey clouds are edged with the soft touch of day. A crow flails and turns in the air above the lake. I tuck down beside the oak, shelter from the wind that drifts in – feeling the rough cracked skin of the bark against my back. I close my eyes and feel the soft touch of the newborn sunlight brushing the cold of night away. The low light of dawn washes over the oak, turning the tree the colour of honey.

From the lake shore, the heron lifts with a graceful elegance. He turns, pale now against the blue sky, and glides away south, following the line of the stream.

ACKNOWLEDGEMENTS

I n the eight years that have seen *The Oak Papers* emerge from some seed of a notion, a handful of people have been vital in supporting the growth to a fully-formed book: my sister Helen Canton, my mother Margaret Canton, Chris and Jude Gibson, Ros Green, Paul Gwynne, Peter Hulme, Jonathan Jukes, Juliet Lockhart, Sara Maitland, my editor Simon Thorogood and my agent Jessica Woollard have each been an essential element in bringing the book to life. I am extremely grateful to them for their various forms of support, guidance, advice and expertise.

There were many others who offered wise words and thoughts or helped me to secure the time and space needed to write the book. I wish to thank them all:

Marigold Atkey, Abdulkareem Atteh, Sarah Beavins, Ronald Blythe, Anna Burton, Sarah Clark, my children Eva, Molly and Joe Canton, Ben Castell, David and Mandy Charleston, Lola Chattam, Katharine Cockin, Richard Cromach, Kim Crowder, Chris Cunningham, Yalda Davis, Katie Dawson, Matthew De Abaitua, Hannah Elliman, Clive Ellis, Richard Fordham, Adrian Gascoyne, Mandy Haggith, Yara Issa, Alex Kimbo, Liz Kuti, Richard Mabey, Matt Mackman, Mark Mansfield, Adrian May, Chris McCully, Ellie Mead, Maria Medlycott, Louise Millar, Edward Milner, Frances Mount, Eliza O'Toole, Andy Papps, Christiana Payne, Jak Peake, Craig Perry, Holly Pester, Clare Pollard, Dylan Pym, Mike Rogerson, Neil Rollinson, Jordan Savage, Mark Self, the staff of the Rare Books and Music Reading Room at the British Library, Fiona Stafford, the students and staff of Honywood Community Science School, Coggeshall (especially Philippe Bilby, Phaedra Bishop, Aidan Tolhurst and Craig Robertson), Stephen Taylor, Phil Terry, Sally Webster, Stephen and Becky Westover, Jules Wilkinson and Jane Winch.

For part of the writing process I was supported by an Arts Council grant which helped enormously and I am very appreciative of their financial funding for the project. There was a wonderful collective of people who agreed to be part of an Oak Papers Reading Group which gave really valuable feedback on early drafts of the book. Thanks to Robyn Bechelet, Ruth Bradshaw, Barbara Claridge, Wendy Constance, Miranda Cichy,

Lelia Ferro, Peter North, Claire Pearson, Stephen Rutt, Molly Shrimpton and Judith Wolton.

To those at Canongate who have worked so hard on the book I also offer my thanks and especially so to Leila Cruickshank, Alice Shortland, Vicki Rutherford and Lucy Zhou. There were also a number of people at Marks Hall Estate who kindly supported me in my time with the Honywood Oak. I would like to particularly thank Ian Chandler, Kath Cockshaw, Sarah Edwards, Rebecca Lee, Elizabeth Pottinger and Richard Ramsey.

To all those other kind folk whose names I have forgotten or never knew but who have also played a part in helping me on this long adventure I send my thanks, too. And, of course, to the oaks.

ENDNOTES

SEEING THE OAK

1 Robert Burton, *The Anatomy of Melancholy* 6th Edition (London: B. Blake, 1838), p.245.

2 *A Choice of Thomas Hardy's Poems*, ed. by Geoffrey Grigson (London: Macmillan, 1969), p. 56.

3 *Kilvert's Diary 1870–1879: Selections from the Diary of The Rev. Francis Kilvert, ed. by William Plomer* (London: Jonathan Cape, 1944), p. 305.

4 William Hayley, *The Life and Letters of William Cowper* (Chichester: W. Mason, 1809), Vol. 4, p. 455.

5 See Michael Tyler, *British Oaks: A Concise Guide* (Marlborough, Wilts: The Crowood Press, 2008), p. 46 on the complex habitats provided by ancient oaks.

6 J.G.D. Clark, *Prehistoric Europe: The Economic Basis* (London: Methuen, 1952), p. 60.

7 See William Bryant Logan, *Oak: The Frame of Civilization* (New York: W.W. Norton, 2005), pp. 46–7.

8 The Writings of Henry David Thoreau, ed. Bradford Torrey, 7 Vols (Boston and New York: Houghton, Mifflin and Co., 1906). I, p.305.

9 *Remaines of Gentilisme and Judaisme*, by John Aubrey, 1686–7; edited and annotated by James Britten (London: W. Satchell, Peyton, and Co., 1881), p. 247. Quoted in James Frazer, *The Golden Bough* [1890], 2nd Ed. (London: Macmillan, 1900), Vol. I, p. 172.

10 Dante Alighieri, *The Divine Comedy*, Part I, 'Hell', Canto XIII, trans. by Dorothy L. Sayers (London: Penguin, 1949), pp. 149–155.

11 William Bryant Logan, *Oak: The Frame of Civilization*, pp. 26–7.

12 Quoted in Frances Carey, *The Tree: Meaning and Myth* (London: British Museum Press, 2012), p. 154.

13 Logan, pp. 86 and 97.

KNOWING THE OAK

1 Pliny, *Natural History*, trans. H. Rackham, 10 vols (London: Heinemann, 1968), IV, pp. 549–551.

2 James Frazer, *The Golden Bough* [1890], 3 vols (London: Macmillan, 1900), III, p. 327.

3 *The Golden Bough*, I, pp. 225–6.

4 Robert Graves, *The White Goddess* (London: Faber, 1961), p. 298.

5 See https://www.poetryfoundation.org/poems/7/lallegro.

6 John Claudius Loudon, *Arboretum et Fruticetum Britannicum, or The Trees and Shrubs of Britain*, Vol. III, Chapter CV [1838] (London: Bohn, 1854), p. 1752.

7 Kathleen Basford, *The Green Man* (Ipswich: D.S. Brewer, 1978), p. 9.

8 Ralph of Coggeshall, *Chronicon Anglicanum*, accessed at http://anomalyinfo.com/Stories/extra-ralph-coggeshalls-account-green-children.

9 Peter Young, *Oak* (London: Reaktion, 2012), p. 117.

10 *The Complete Works of John Keats*, ed. H. Buxton Forman, Vol. II (New York: Thomas Crowell, 1900), p. 192.

11 Thomas Keightley, *The Fairy Mythology*, illustrative of the romance and superstition of various countries. [1828] (London: H.G. Bohn, 1870), pp. 290–1.

12 On Puck as Robin Goodfellow see William Shakespeare, *A Midsummer Night's Dream*, ed. Peter Holland (Oxford: Oxford University Press, 1994), 'Introduction', pp. 35–49.

13 Rudyard Kipling, *Puck of Pook's Hill* [1906], The Writings in Prose

and Verse of Rudyard Kipling, Vol. XXIV Edition de Luxe (London: Macmillan, 1907), p. 11.

14 Elsie Innes, *The Elfin Oak of Kensington Gardens* (London: Frederick Warne, 1930), n.p.

15 D.H. Lawrence, 'Under the Oak', in *New Poems*, (London: Martin Secker, 1918). See https://www.bartleby.com/128/14.html.

16 Robert Graves, *The White Goddess* (London: Faber, 1961), p. 298.

17 William Shakespeare, *As You Like It*, ed. Alan Brissenden (Oxford: Oxford University Press, 1993), 4, iii, 105–8, (p. 203).

18 See http://greatpoetryexplained.blogspot.com/2019/01/domicilium -by-thomas-hardy.html.

19 Michael Tyler, *British Oaks: A Concise Guide* (Marlborough, Wilts.: The Crowood Press, 2008), p. 92.

20 Jacob George Strutt, *Sylva Britannica; or Portraits of Forest Trees, distinguished for their antiquity, magnitude or beauty* (London: A.J. Valpy, 1822), p. 1.

21 Strutt, *Sylva Britannica*, p. 10.

22 William Gilpin, *Remarks on Forest Scenery and Other Woodland Views*, 2 vols (London: R. Blamire, 1791), p. 27.

23 Joseph Taylor, *Arbores Mirabiles: or a description of the most remarkable trees, plants and shrubs, in all parts of the world* (London: W. Darton, 1812), pp. 106 and 108.

24 Peter Young, *Oak* (London: Reaktion, 2012), pp. 70 and 72.

25 Gilbert White, *Gilbert White's Journals*, ed. Walter Johnson, (Newton Abbot, Devon: David & Charles, 1970), p. 261.

26 Tyler, p. 182.

27 Louis MacNeice, *Collected Poems* (London: Faber, 2007), p. 272.

28 John Clare, *Poems Descriptive of Rural Life and Scenery* (London: Taylor and Hessey, 1820), p. 208.

29 *The Works of the British Poets*, ed. Robert Anderson (London: J. & A. Arch, 1795), Vol. 6, p. 242.

30 J.C. Shenstone, *The Oak Tree in Essex* (n.p: n. pub, 1894). The booklet is apparently taken from a paper read at Essex Field Club on 23 June 1894.

31 W.H. Davies, *Collected Poems* (London: Jonathan Cape, 1943), p. 174.

32 *Collected Poems*, p. 140.

33 John Fowles, *The Tree* (St Alans, Herts: The Sumach Press, 1992), p. 31.

34 Thomas Hardy, *Far from the Madding Crowd*, ed. Ronald Blythe (London: Penguin, 1978), p. 58.

35 Edmund Spenser, *The Shepheardes Calender* (1579) quoted in John Claudius Loudon, *Arboretum et Fruticetum Britannicum, or The Trees and Shrubs of Britain*, Volume III, Chapter CV, [1838] (London: Bohn, 1854), p. 1785.

36 *The Eclogues of Virgil*, trans. C. Day Lewis (London: Jonathan Cape, 1963), p. 33.

BEING WITH OAKS

1 Martin Buber, *I and Thou*, trans. Walter Kaufmann (Edinburgh: T & T Clark, 1970), pp. 57–9.

2 Gary Snyder, 'Kitkitdizze: A Node in the Net', in *A Place in Space: Ethics, Aesthetics and Watersheds* (Washington D.C.: Counterpoint, 1995), p. 263.

3 Ralph of Coggeshall, *Chronicon Anglicanum*, quoted in Thomas Keightley, *The Fairy Mythology* (London: H. G. Bohn, 1870), pp. 281–2.

4 Daniel Defoe, *Mere Nature Delineated: Or, a Body without a Soul: Being Observations upon the Young Forester lately brought to town from Germany* (London: T. Warner, 1726), pp. 5 and 16.

5 Henry Wilson, *Wonderful Characters*, Vol II (London: Robins, 1821), pp. 152–160.

6 Henry David Thoreau, 'The Writings of Henry D. Thoreau, Journal Volume 2: 1842–1848', ed. Robert Sattelmeyer (Princeton: Princeton University Press, 1984), p. 37.

7 Virginia Woolf, *Orlando* (Edinburgh: Canongate, 2012), pp. 5–6.

8 See Italo Calvino, 'The Baron in the Trees', trans. Archibald Colquhoun in *Our Ancestors* (London: Picador, 1980) .

9   *Far from the Madding Crowd*, p. 88.

10 Leo Tolstoy, *War and Peace*, trans. Louise and Aylmer Maude [1868] (London: Macmillan, 1943), pp. 454–9.

11 Stephen Taylor, *Oak: One Tree, Three Years, Fifty Paintings* (New York: Princeton Architectural Press, 2012), p. 56.

12 T.S. Eliot, 'Burnt Norton', in *Four Quartets* (London: Faber, 1963), p. 13, I. 19.

13 William Shakespeare, *The Merry Wives of Windsor*, ed. Giorgio Melchiori (London: Thompson, 2000), 4, iv, 26–9 (p. 257).

14 Q. Li et al., 'Effect of phytoncide from trees on human natural killer cell function', *International Journal of Immunopathology and Pharmacology*. Oct.–Dec. 2009, 22(4) pp. 951–9.

15 See www.monicagagliano.com.

16 Monica Gagliano, 'The Mind of Plants: Thinking the Unthinkable', *Communicative & Integrative Biology* (2017) 10:2.

17 Paco Calvo, 'The Philosophy of Plant Neurobiology: A Manifesto', *Minimal Intelligence Lab* (MINT Lab), *Synthese* (2016) 193: 1323.

18 Suzanne Simard, 'Note from a Forest Scientist' in Peter Wohlleben, *The Hidden Life of Trees* (Vancouver: Greystone Books, 2016), p. 249.

19 Stuart Piggott, *The Druids* (London: Thames and Hudson, 1968), pp.15–16.

SELECT BIBLIOGRAPHY

Anderson, Robert, ed., *The Works of the British Poets* (London: J. & A. Arch, 1795)

Aubrey, John, *Remaines of Gentilisme and Judaisme*, ed. James Britten (London: W. Satchell, Peyton, and Co., 1881)

Baker, J.A., *The Peregrine* (London: William Collins, 1967)

Basford, Kathleen, *The Green Man* (Ipswich: D. S. Brewer, 1978)

Buber, Martin, *I and Thou*, trans. Walter Kaufmann (Edinburgh: T & T Clark, 1970)

Burton, Robert, *The Anatomy of Melancholy* 6th Edition (London: B. Blake, 1838)

Calvino, Italo, 'The Baron in the Trees', trans. Archibald Colquhoun in *Our Ancestors* (London: Picador, 1980)

Calvo, Paco, 'The Philosophy of Plant Neurobiology: A Manifesto', *Minimal Intelligence Lab* (MINT Lab), *Synthese* (2016) 193: 1323

Carey, Frances, *The Tree: Meaning and Myth* (London: British Museum Press, 2012)

Clare, John, *Poems Descriptive of Rural Life and Scenery* (London: Taylor and Hessey, 1820)

Clark, J.G.D., *Prehistoric Europe: The Economic Basis* (London: Methuen, 1952)

Dante (Alighieri), *The Divine Comedy*, Part I, 'Hell', Canto XIII, trans. Dorothy L. Sayers (London: Penguin, 1949)

Davies, W.H., *Collected Poems* (London: Jonathan Cape, 1943)

Defoe, Daniel, *Mere Nature Delineated: Or, a Body without a Soul: Being*

Observations upon the Young Forester lately brought to town from Germany (London: T. Warner, 1726)

Eliot, T.S., *Four Quartets* (London: Faber, 1963)

Forman, H. Buxton, ed., *The Complete Works of John Keats* (New York: Thomas Crowell, 1900)

Fowles, John, *The Tree* (St Albans, Herts: The Sumach Press, 1992)

Frazer, James, *The Golden Bough*, 2nd ed. (London: Macmillan, 1900)

Gagliano, Monica, 'The Mind of Plants: Thinking the Unthinkable', *Communicative & Integrative Biology* (2017) 10:2

Gilpin, William, *Remarks on Forest Scenery, and Other Woodland Views*, 2 vols (London: R. Blamire, 1791)

Graves, Robert, *The White Goddess* (London: Faber, 1961)

Grigson, Geoffrey, ed., *A Choice of Thomas Hardy's Poems* (London: Macmillan, 1969)

Hardy, Thomas, *Far from the Madding Crowd*, ed. Ronald Blythe (London: Penguin, 1978)

Hayley, William, ed., *The Life and Letters of William Cowper* (Chichester: W. Mason, 1809)

Innes, Elsie, *The Elfin Oak of Kensington Gardens* (London: Frederick Warne, 1930)

Johnson, Walter, ed., *Gilbert White's Journals* (Newton Abbot, Devon: David & Charles, 1970)

Keightley, Thomas, *The Fairy Mythology* (London: H.G. Bohn, 1870)

Kipling, Rudyard, 'The Writings in Prose and Verse of Rudyard Kipling', Vol. XXIV, Edition de Luxe (London: Macmillan, 1907)

Lawrence, D.H., *New Poems* (London: Martin Secker, 1918)

Li, Q., et al., 'Effect of phytoncide from trees on human natural killer cell function', *International Journal of Immunopathology and Pharmacology*. Oct.–Dec. 2009, 22(4)

Logan, William Bryant, *Oak: The Frame of Civilization* (New York: W.W. Norton, 2005)

Loudon, John Claudius, *Arboretum et Fruticetum Britannicum, or The Trees and Shrubs of Britain*, (London: Bohn, 1854)

MacNeice, Louis, *Collected Poems* (London: Faber, 2007)

Piggott, Stuart, *The Druids* (London: Thames and Hudson, 1968)

Pliny, *Natural History*, trans. H. Rackham, 10 vols (London: Heinemann, 1968)

Plomer, William, ed., *Kilvert's Diary 1870–1879: Selections from the Diary of The Rev. Francis Kilvert* (London: Jonathan Cape, 1944)

Ralph of Coggeshall, *Chronicon Anglicanum*

Sattelmeyer, Robert, ed., 'The Writings of Henry D. Thoreau: Journal Volume 2: 1842–1848' (Princeton: Princeton University Press, 1984)

Shakespeare, William, *A Midsummer Night's Dream*, ed. Peter Holland (Oxford: Oxford University Press, 1994)

———, *As You Like It*, ed. Alan Brissenden (Oxford: Oxford University Press, 1993)

———, *The Merry Wives of Windsor*, ed. Giorgio Melchiori (London: Thompson, 2000)

Shenstone, J.C., *The Oak Tree in Essex* (n.p: n. pub, 1894)

Snyder, Gary, *A Place in Space: Ethics, Aesthetics and Watersheds* (Washington D.C.: Counterpoint, 1995)

Strutt, Jacob George, *Sylva Britannica* (London: A.J. Valpy, 1822)

Taylor, Joseph, *Arbores Mirabiles* (London: W. Darton, 1812)

Taylor, Stephen, *Oak: One Tree, Three Years, Fifty Paintings* (New York: Princeton Architectural Press, 2012)

Tolstoy, Leo, *War and Peace*, trans. Louise and Aylmer Maude (London: Macmillan, 1943)

Tyler, Michael, *British Oaks: A Concise Guide* (Marlborough, Wilts: The Crowood Press, 2008)

Virgil, *The Eclogues*, trans. C. Day Lewis (London: Jonathan Cape, 1963)

Wilson, Henry, *Wonderful Characters* (London: Robins, 1821)

Wohlleben, Peter, *The Hidden Life of Trees* (Vancouver: Greystone Books, 2016)

Woolf, Virginia, *Orlando* (Edinburgh: Canongate, 2012)

Young, Peter, *Oak* (London: Reaktion, 2012)

ABOUT THE AUTHOR

JAMES CANTON has taught the MA course in Wild Writing at the University of Essex since its inception in 2009, exploring the ties between literature, landscape, and the environment. He is the author of *Ancient Wonderings: Journeys Into Prehistoric Britain* and *Out of Essex: Re-Imagining a Literary Landscape*. He has written reviews for *The Times Literary Supplement*, *Caught by the River*, and *EarthLines*. He has also appeared on television and radio and regularly gives talks and workshops.